Synthesis Lectures on Chemical Engineering and Biochemical Engineering

This series publishes short books on all aspects of chemical engineering, covering the analysis or design of chemical processes to effectively convert materials into more useful materials or energy. The books will focus on fundamental aspects necessary for chemical engineering design including chemistry, math, physics, and sometimes biology to improve the quality of life by inventing, optimizing, and economizing new technologies and products.

Javid A. Parray · Wen-Jun Li

Microbial and Enzyme-Based Technology for Plastic Biodegradation

Springer

Javid A. Parray
State Key Laboratory of Desert and Oasis
Ecology, Xinjiang Institute of Ecology
and Geography
Chinese Academy of Sciences
Urumqi, China

Department of Environmental Science
GDC Eidgah
Cluster University Srinagar
Srinagar, Jammu and Kashmir, India

Wen-Jun Li
School of Life Sciences
Sun Yat-sen University
Guangzhou, China

ISSN 2327-6738   ISSN 2327-6746   (electronic)
Synthesis Lectures on Chemical Engineering and Biochemical Engineering
ISBN 978-3-031-84436-2   ISBN 978-3-031-84437-9   (eBook)
https://doi.org/10.1007/978-3-031-84437-9

Acknowledgment

I am writing to express our sincere and profound gratitude to the Director of the Xinjiang Institute of Ecology and Geography, Chinese Academy of Science, Xinxiang, Urumqi, China, for their support under the CAS President's International Fellowship Initiative (PIFI) for Project number—2024PVB0057 Dr. Javid A. Parray under the supervision of Prof. (Dr.) Wen-Jun Li. The Key-Area Research and Development Program of Guangdong Province (2022B0202110001) also supported this work. Finally, we are grateful to Prof. Li and his team, Dr. Yong-Hong Liu, Dr. Osama Mohamad, Dr. Wasim Sajjad, Dr. Rashidin Abdugheni and Associate Profs. Bao-Zhu Fang and Suai Li, and Murad Muhammad. Their unwavering encouragement and support were instrumental in completing our study. We would also like to extend our heartfelt appreciation to the administrative staff of the Xinjiang Institute, whose often behind-the-scenes efforts were crucial in facilitating our research process. Their efficiency in managing logistical aspects considerably eased my journey and allowed us to focus on the core objectives of our project.

Moreover, we acknowledge the importance of the collaborative spirit evident within the institute, which fostered an environment ripe for learning and innovation. The interactions and discussions with fellow researchers enriched our understanding and inspired new perspectives that enhanced the overall quality of our findings. The proactive sharing of knowledge by all team members exemplifies a commitment to academic excellence that is truly commendable. As we reflect on this project's journey, we recognize the profound impact of mentorship and guidance on our academic pursuits. Professor Li's insightful feedback and encouraging words at every stage motivated us to strive for excellence, reminding us of the importance of perseverance in research.

In conclusion, this collaborative effort is a testament to the power of partnership and collective endeavour within the scientific community. We look forward to future collaborations and hope to contribute to the ongoing quest for knowledge and advancement in our respective fields. Our shared commitment to scientific inquiry will undoubtedly pave the way for even more outstanding achievements in the future.

Javid A. Parray
PIFI–CAS fellow

Contents

About the Authors

Dr. Javid A. Parray holds a Master's Degree in Environmental Science. He completed his M.Phil. and Ph.D. research programs at the University of Kashmir after passing the prestigious state-level JKSLET examination. He also conducted post-doctoral research at the same university and was awarded a Fast Track Young Scientist Project by the SERB-DST, Government of India, New Delhi. Dr. Parray teaches in the Department of Environmental Science at GDC Eidgah Srinagar, part of the Cluster University Srinagar. He has participated in courses and conferences on environmental and biotechnological aspects in countries such as Sri Lanka, Indonesia, Malaysia, China, and Uzbekistan. His research interests include ecological and agricultural microbiology, climate change and microbial biotechnology, environmental microbiomes, environmental sustainability, and microbial genomics. Dr. Parray has published over 50 high-impact research papers and book chapters in esteemed journals and publishing platforms. Additionally, he has authored 30 books with international publishers, including Elsevier, Springer, Callisto Reference USA, and Wiley-Blackwell. Dr. Parray serves on editorial boards, is a permanent journal reviewer, and has been invited to speak at various scientific meetings and conferences in India and abroad. He has also guest-edited special issues on Environmental Biofilms and Sustainable Food Systems with BioMed Research International, Hindawi, and Frontiers in Sustainable Food Systems. Dr. Parray is a member of several international and national scientific organizations, including the Asian PGPR Society, IJMS Mumbai,

Academy of Eco Science, and IAES Haridwar. In recognition of his work, he was awarded the "Emerging Scientist Year Gold Medal" for 2018 by the Indian Academy of Environmental Science. Furthermore, Dr. Parray serves as the Course Coordinator for three CeC-MOOCS National Swayam courses in Environmental Science and has received a PIFI Visiting Fellowship from the Chinese Academy of Science. He is also the series editor of two book series focusing on Microbiome Research in Plants and Soil and Microbial Genomics.

Prof. Dr. Wen-Jun Li received his Ph.D. in Microbiology from Shenyang Institute of Applied Ecology, Chinese Academy of Sciences, in 2002. He is currently working as Distinguished Professor in School of Life Sciences, Sun Yat-Sen University, Guangzhou, China. His publications include eight monographs, 30 authorized patents, and more than 700 peer-reviewed papers (https://www.researchgate.net/profile/Wen-Jun-Li) as the first or corresponding authors. His research mainly focuses on microbial diversity under those unusual environments, by using culture-dependent and culture-independent methods, and also on microbial ecology and evolution. He was awarded the World Federation for Culture Collections (WFCC) Skerman award for microbial taxonomy, and other 10 leading awards for his outstanding research contributions on the field of microbial systematics and microbial ecology.

He is serving as editorial board members or Associate Editors of more than 10 International journals, and he was appointed as membership of International Committee on Systematics of Prokaryotes (ICSP) since 2017, and President of Bergey's International Society for Microbial Systematics (BISMiS, https://www.bismis.net/bismislive.html) since December of 2023 till date.

Email: liwenjun3@mail.sysu.edu.cn

http://lifesciences.sysu.edu.cn/teachers/professor/243

https://www.researchgate.net/profile/Wen_Jun_Li/publications

Introduction to Plastic Waste: A Growing Global Concern

1.1   Introduction

Humans have gone through several eras, some of which—such as the Stone Age, Bronze Age, and Iron Age—have been named for the materials used to make amazing tools and other essentials. Since we are currently living in the Plastic Age, history is being repeated. The Greek term "Plasticos," which means "to be able to form," is where the word "plastic" originated. The synthetic polymer known as plastic is created by polymerizing monomers that are taken from petrochemicals and mixed with other substances [1]. Long-chain polymers consisting of carbon, hydrogen, and oxygen bound together by covalent bonds are called monomers. Lightweight monomers like ethylene and propylene are frequently utilized in the production of plastics [2]. Certain chemical groups may be left off of each monomer throughout the polymerization process, which causes polymers to varying degrees retain the reactivity or chemical characteristics of the original monomer unit. There are two types of polymerization: "addition," which involves joining monomers while preserving their original structure, and "condensation," which involves changing the structure of monomers and producing by products like water. Depending on the kind of plastic produced or the manufacturing process, different chemicals are employed in different plastics. Fillers, plasticizers, pigments, foaming agents, processing aids, lubricants, heat stabilizers, acid scavengers, antioxidants, UV stabilizers, flame retardants, and some antistatic agents are among the additives used in the production of plastic in varying amounts to control various functions such as improving properties, ease of processing, durability, performance, and appearance. Biomaterials are currently being extensively investigated by the industrial sectors as potential substitutes for conventional plastic materials derived from petroleum [3]. Natural resources like vegetable oils and starches (plant-based products) make up all or part of these bioplastics. Bioplastics have

J. A. Parray and W.-J. Li, *Microbial and Enzyme-Based Technology for Plastic Biodegradation*, Synthesis Lectures on Chemical Engineering and Biochemical Engineering, https://doi.org/10.1007/978-3-031-84437-9_1

the potential to lessen their negative effects on the environment, particularly their carbon footprint and the greenhouse gas emissions linked to the usage of polymers. The goal of reducing landfill trash propelled the creation of contemporary bioplastics in the 1980s [4]. Consequently, the main goal of bioplastic production is to produce an eco-friendly product that is on par with traditional plastics in terms of quality. By 2025, the global market for bioplastics and biopolymers is expected to grow from USD 10.5 billion in 2020 to USD 27.9 billion, despite the fact that bioplastics only make up 1% of all plastic manufacturing [5]. Among the many beneficial qualities of plastics are their low weight, hardness, high strength, good stiffness, versatility in manufacturing and design, good insulation, poor electrical and thermal conductivity, and resistance to corrosion [6]. Excellent physio-chemical qualities, including high heat distortion temperatures, exceptional mechanical strengths, superior chemical resistance, and electrical insulation, are possessed by synthetic semiaromatic polymers, such as polyethylene terephthalate (PET) and poly butylene terephthalate (PBT). Their capacity to retain strength and shape at high temperatures is demonstrated by characteristics like their high melting temperature (Tm) and glass transition temperature (Tg). Plastic is becoming a vital component of every human life due to its low production costs, malleability, and processing capabilities. As a result, those qualities make polymers more useful, particularly during World War II when nylon was developed. Plastics have a wide range of applications in manufacturing, electronics, furniture, toys and leisure products, transportation, construction, agriculture, packing, medicine, and heat insulation [6]. Because plastics are lightweight and biocompatible, they are currently used in the production of 85% of medical devices, including joint replacement, sterile packaging, disposable syringes, and intravenous bags [7]. The benefits of plastics can be divided into three categories: (i) environmental (e.g., PET beverage packaging materials reduce energy consumption and greenhouse gas emissions compared to glass and metal packing, and lighter plastic composites in aircraft result in low fuel consumption and easy assembly), (ii) social (e.g., by installing a variety of water controlling and distribution systems, clean drinking water can be supplied and stored, and food packaging allows for the safe and time-depend distribution. Due to a lack of raw materials in the current day, plastics are gradually replacing traditional materials including glass, wood, metal, and paper [8]. Consequently, the current pattern of plastic consumption has led to an increase in worldwide plastic production [9]. Furthermore, at the current rate of expansion, plastic manufacturing is predicted to treble over the next 20 years [10]. In 2010, China overtook Europe as the world's top manufacturer of plastic. China continues to be the world's biggest producer and consumer of plastic, accounting for 60 MT of the 359 MT produced in 2018. Additionally, one-third of the world's plastics are produced in China [6]. In contrast to Asian nations, Western nations are thought to be the largest per capita consumers.

1.2   Thermoplastic Polymers

Since thermoplastic polymers (TP) are inexpensive, have low moulding temperatures, are lightweight, and have a variety of chemical, thermal, and optical characteristics, they are most frequently used as a work piece to create micro-patterns using the hot embossing technique. Amorphous and semi-crystalline polymers are the two categories into which thermoplastic polymers fall. According to their thermal stability, each category is further divided into three categories: standard, technical, and high-performance polymers [11]. Figure 1.1 displays the comprehensive categorization of thermoplastic polymers according to their thermal stability. Transparency is a crucial characteristic for optical applications such as light-guided plates, Fresnel lenses, and micro-lenses. When compared to an amorphous polymer, the crystalline portion of a semi-crystalline polymer generates more refraction, which reduces transparency. Compared to semi-crystalline polymers, amorphous polymers such as polymethyl methacrylate and polycarbonate are best suited for optical applications. Because of their exceptional chemical resistance, semi-crystalline polymers are perfect for producing microfluidic devices for both chemical and medical purposes. This mould may not be reusable due to its high chemical resistance; therefore, it is essential to address the mould coating and set of operating parameters at an optimal level. If the polymer is still present in the micro-cavities (forming the patterns) on the mould, chemical cleaning will be difficult. The shrinkage that occurs in semi-crystalline polymers during the hot embossing process can be divided into two parts: one that occurs perpendicular to the molecular orientation and one that occurs along the molecular orientation. In contrast to amorphous polymers, this shrinkage causes non-uniform shrinkage in semi-crystalline polymers and makes it challenging to forecast the shrinkage in semi-crystalline polymers. Depending on the mould design, polymer type, and aspect ratio of the micro-patterns to be embossed, the amorphous polymer embossing temperature range is from 30 to 50 °C. On the other hand, the semi-crystalline polymer's embossing temperature range is from 3 to 5 °C, therefore choosing the right temperature range is essential for successful embossing [11]. Visco elastic bodies, a combination of an elastic body and a viscous fluid, include thermoplastic polymers. The deformation behaviour is explained by the modulus of the polymer workpiece, which varies with time and temperature. Deformation is classified into three states based on the modulus measured at different temperatures (deformation of thermoplastic polymer with respect to temperature) and time periods (deformation of thermoplastic polymer with respect to time). In the glassy state, the polymer substrate is distorted when Te is smaller than Tg. This fully elastic deformation results from the lengthening of the atomic distance. As temperature increases, the substrate reaches a rubber-like condition where it becomes soft and behaves like incompressible rubber [12].

The localized movement of the polymer chain is what causes the deformation of the polymer in the rubbery state. Its modulus is greatly decreased as a result, making it easier for the polymer substrate to deform. The thermal energy is enough to overcome

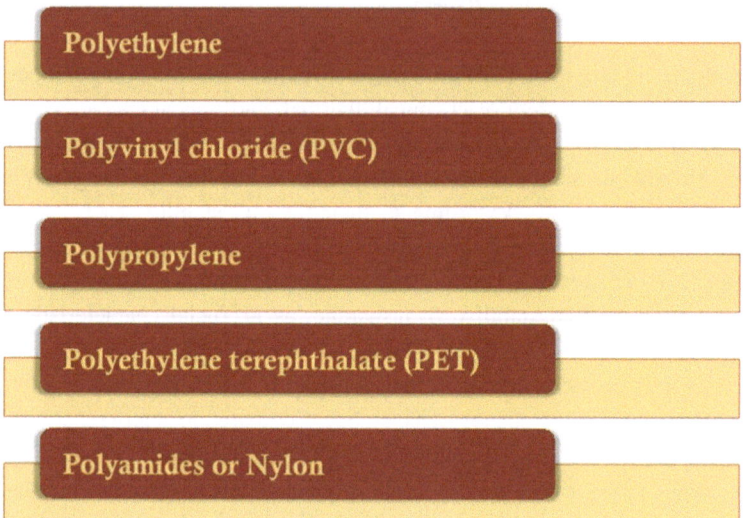

Fig. 1.1 Common forms of synthetic plastics

the potential barrier for the rotational and translational movement of polymer molecules, and the amplitude of vibrational motion of the molecules becomes more important as the temperature approaches Tg. The ability of segments of polymer molecules to freely migrate between lattices makes the material robust. This makes it easier to deform in a rubbery condition as opposed to a glassy one [13]. However, parts of polymer chains can move locally and reversibly. Small deformation is seen in the rubbery condition after sufficient processing time, and the polymer tends to regain the distortion when the load is released. To prevent any flaws, the work item is heated above Tg and below Tm during the embossing process. This indicates that embossing is done in a rubbery state. Thus, this recoverable deformation in the rubbery condition has a major impact on the precision of the embossed component. The polymer substrate's modulus and viscosity decrease with increasing temperature, and it transitions into the viscous flow state when subjected to external loading. In this state, chain sliding occurs, which causes movement in the polymer chain, and it flows. The deformation obtained after this is permanent and not reversible, unlike the deformation in a rubbery state.

1.3   Global Perspectives

Scientists from every continent have seen the pervasiveness of plastic pollution, which is a defining feature of contemporary culture. Researchers have discovered plastic particles on the summit of Mount Everest, in rainwater [14], in the deep sea [15], and even in human blood and placenta [16]. Every year, it chokes or starves millions of birds, damages innumerable sea creatures, and is present in the water we drink and consumed by plankton [17]. According to a recent analysis, plastics are definitely having an impact on the environment, but it is much harder to determine how much damage microplastics are inflicting [18]. It makes sense in light of this situation to conclude that plastic pollution directly impedes global efforts for sustainable development [19]. One important indicator of the Anthropocene may be the pervasiveness of plastic contamination in our landfills, water systems, and even our own bodies. Scientists have called for plastic density and other metrics to be added to the list of factors required to determine our proximity to planetary boundaries [20], and terms like "our plastic age," "plastiglomerate," "anthropoquinas," and "the plastisphere" [21] have emerged. The problem's infinite scope makes ocean and coastal governance more complicated at the municipal, national, and regional levels [22]. Evidence indicates that plastic pollution is getting worse, even though it has received more attention in the past ten years. In contrast to the short-lived decrease in air pollution levels brought on by the COVID-19 epidemic, there is probably more plastic waste in freshwater and marine environments. In fact, attempts to reduce the production and usage of plastic have suffered greatly as a result of the pandemic. Despite the general increase in prohibitions and/or tax-based incentives to reduce use, it has revitalized the single-use plastic industry. However, there was already a call for a significant change in the global cycle of production, consumption, and waste before the epidemic. The absence of coordination among several public and commercial players, as well as the lack of coordination between local and global levels of governance, are some of the reasons why management solutions for litter leakage continue to confront obstacles and exhibit inefficiency [23]. Distinct nations may experience distinct stages of the plastic lifecycle, ranging from the extraction of oil to the production, use, and disposal of products. It is evident that the detrimental effects of marine plastic pollution have a spillover effect that transcends national boundaries and must be tracked. The transboundary nature of this issue necessitates international solutions and diplomatic agreements, even though waste management policies (including recycling and safe disposal) are under national jurisdiction. Many aspects of both production and disposal cross geopolitical borders. The origin and fate of plastic pollution have changed significantly over the past few decades [24], and it is undeniable that a large portion of it ends up in the Global South's canals, riverways, and shorelines, where people immediately suffer the effects and sustainable development becomes more challenging. Although the problem of plastic pollution is not only environmental or economic, it is central to modern environmental justice; a United Nations report highlights that, similar to climate justice, plastic justice has become a prominent concern in the modern era [25]. In order to address plastic pollution, distributive justice issues must be taken into consideration just as much as technical limitations. Furthermore, it is critical that plastic pollution be taken into account when making commitments to sustainable development, emphasizing how this worldwide issue affects people differently and the necessity of identifying development opportunities that can advance environmental justice. Nevertheless, there are still major barriers to putting effective measures to lessen the issue into place, even in light of the growing awareness of the effects of plastic pollution and the necessity of better monitoring systems, regulations, and multisectoral approaches. Due to these challenges, which include international environmental governance, plastic pollution is now regarded as a "wicked problem," especially in the maritime environment [26]. Innovative approaches are required, and they call for a comprehensive knowledge of plastic pollution, especially of the underlying reasons of the current situation and the function of international sustainability agendas in resolving the problem. Therefore, the goal of this research is to provide an exploratory review in order to examine the connections between plastic pollution and sustainable development in this particular setting. Since marine litter directly impacts billions of people who live in coastal communities and the ocean serves as the primary worldwide sink for plastic pollution, it is the primary focus of our attention. However, there are also very encouraging prospects going forward that suggest the novel theme of plastic justice, in conjunction with efforts to combat climate change, biodiversity loss, and ecoviolence [27], can serve to unite communities in empowering initiatives. Marine litter and plastic pollution in particular are negatively impacting the prospects for sustainable development throughout the Global South. We first analyze the issue of plastic pollution from the standpoint of the Sustainable Development Goals (SDGs) and the North–South dynamics of a global plastic waste business before providing a brief case study from Brazil to highlight this tendency. We then present the idea of plastic justice as a progressive normative design and highlight important research needs and gaps.

Few studies have attempted to expressly address how plastic pollution intersects with the SDGs, despite the fact that there is a growing body of information on many different elements of the problem (such as the effects on the environment and human health, alternative materials, waste management techniques, and the circular economy). The 17 SDGs were introduced by the UN in 2015 as a component of the 2030 Agenda for Sustainable Development [28]. The SDGs aim to address shared issues that impact economic, social, and environmental aspects in order to focus international efforts towards a sustainable future. Poverty, hunger, human health, education, gender equality, water resources, energy, work and economy, industry, inequality, cities and communities, consumption and production, climate, marine and terrestrial life, peace and justice, and the collaborations required to promote these interconnected issues are all included in this. All of the SDGs have obvious connections, and the 2030 Agenda places a strong emphasis on comprehending and addressing these connections in the policy domains of each SDG [29]. 231 distinct indicators (as well as a dozen indicators that are repeated across a few SDGs) have been proposed to support the 169 targets linked to the 17 goals in this ambitious project [19]. Plastic pollution is indicated by only one original indicator (item 14.1.1.b: density of floating plastic debris >2.5 cm) [30]. Aiming to "prevent and significantly reduce marine pollution of all kinds, in particular from land based activities, including marine debris and nutrient pollution," goal 14.1 includes this indicator. However, this indicator does not include microplastics (or smaller particles) or objects that are not buoyant (such as weighed-down fishing nets, metals, and wet fabrics, among others) because it only considers plastic that is greater than 2.5 cm. This is a serious constraint since microplastic and nanoplastic pose a threat to the ecosystem services that billions of people depend on. This is especially true in the context of the Global South, where mariculture is used to provide food. Economically deprived people tend to build more nature-based livelihoods, so they are naturally dependent on environmental quality for their well-being, even though the valuations of coastal and marine ecosystem services are proportionately higher in countries with higher HDI and GDP (perhaps due to methodological approaches) [31]. Given the urgency of the issue, 14.1.1.b was intended to be completed by 2025 rather than 2030, as is the case with most milestones. Regretfully, the most recent UN SDG report makes no reference to the progress made in reducing marine waste. Although it is obvious that we must address components all along this chain, in a systemic approach, the fact that the only indicator within the SDG framework that is directly related to plastic pollution is established at the very end of the production–consumption–waste chain of events may be even more concerning than the lack of reliable indicators for marine litter. The most recent UN SDG report has not reported on the progress of two other indirect indicators that are pertinent to the issue of plastic pollution: 11.6.1 ("proportion of municipal solid waste collected and managed in controlled facilities out of total municipal waste generated, by cities") and 12.5.1 ("national recycling rate, tonnes of material recycled"). The connections between SDGs 1 (no poverty), 2 (zero hunger), 8 (decent work and economic growth), 11 (sustainable cities and communities), 12 (responsible consumption and production), and 13 (climate action) were assessed in a report published by the International Council for Science (ICSU, 2017). The research identified 10 objectives that are closely related to target 14.1, all of which would co-benefit from integration, despite the fact that it does not methodically distinguish plastic waste from other types of marine pollution. Targets and goals can be accomplished more effectively or at a reduced cost by recognizing co-benefits and coordinating actions accordingly [29]. In keeping with the urgency of the issue, the report also notes that, in contrast to other targets, 14.1.1.b was intended to be completed by 2025 [32]. Regretfully, the most recent UN SDG report makes no reference to the progress made in reducing marine waste [33]. Although it is obvious that we must address components all along this chain, in a systemic approach, the fact that the only indicator within the SDG framework that is directly related to plastic pollution is established at the very end of the production–consumption–waste chain of events may be even more concerning than the lack of reliable indicators for marine litter. The most recent UN SDG report [33] has not reported on the progress of two other indirect indicators that are pertinent to the issue of plastic pollution: 11.6.1 ("proportion of municipal solid waste collected and managed in controlled facilities out of total municipal waste generated, by cities") and 12.5.1 ("national recycling rate, tonnes of material recycled"). SDGs 1 (no poverty), 2 (zero hunger), 8 (decent work and economic growth), 11 (sustainable cities and communities), 12 (responsible consumption and production), and 13 (climate action) were all linked to SDG 14 (life below water), according to a report by the International Council for Science (ICSU, 2017). The research identified 10 objectives that are closely related to target 14.1, all of which would co-benefit from integration, despite the fact that it does not methodically distinguish plastic waste from other types of marine pollution. Targets and goals can be accomplished more effectively or at a reduced cost by recognizing co-benefits and coordinating actions accordingly [29].

1.4   Environmental Impacts

Due to poor management, ineffective waste management strategies, and their ongoing release, the globe currently generates an astounding amount of plastic garbage. By the end of 2015, an estimated 5800 Mt of improperly managed plastic garbage had been dumped into the environment worldwide [34]. Because of their stability, endurance, resistance to biodegradation, large molecular weight, intricate three-dimensional structure, and hydrophobic nature, these plastics persist in the environment for many years, perhaps even centuries [35]. In the environment, plastics can weather through a variety of processes, including fragmentation, hydrolysis, photodegradation, thermal oxidation, and biodegradation. However, plastics are found in a wide range of environments, including farms, deserts, mountaintops, and oceans [36].

1.5   Plastics in Terrestrial and Marine Environments

Waste builds up in the terrestrial environment because plastics are mostly produced, consumed, and disposed of on land. Poly tunnel plastic films, packing materials, and wrappings are examples of agricultural plastics that are widely available as plastic trash. These plastic particles can linger in the soil for up to 15 years after breaking down into microplastics. Furthermore, the amount of plastic that accumulates on land is increased by unlawful dumping and direct garbage input [37]. They can be exposed to a certain extent by the terrestrial environment, which can serve as a plastic sink. Plastic litter is carried by human/anthropogenic, wind (aeolian), and water (fluvial) activity. Plastic waste has overtaken freshwater systems, including rivers, lakes, dams, and urban drainage networks, in addition to land [38]. Because of their natural location in valleys and lower elevation areas, such as floodplains, freshwater habitats are the main container for different contaminants discharged within a watershed [38]. One of the main ways that plastics enter the marine environment is through rivers [37]. The United Nations Environmental Programme (UNEP) estimates that approximately 1,000 rivers contribute between 0.8 and 2.7 million tonnes of riverine plastic emissions annually, or nearly 80% of global riverine plastic emissions. In cities, wind is the primary means of moving tiny plastic particles, or micro- and nanoplastics. Micro- and nanoplastics can be dispersed across a large area via atmospheric circulations [38]. As a result, the mobilization and accumulation of plastics in the atmosphere are significantly influenced by particle size. According to [39], three main factors—watershed outfalls, population density, and marine activity—determine how much plastic enters the ocean. Eighty per cent of all marine pollution is caused by the 8 to 10 million tonnes of plastics, including both macro- and microplastics, that flow into the ocean each year. Over 100,000 marine species are killed annually by plastic garbage, which affects everything from beaches to the deep ocean [40]. Tonnes of plastic debris are dumped into the Indian Ocean by maritime mishaps such as the MV X-Press Pearl tragedy [41]. Plastic garbage can spread across a wide area because of buoyancy, durability, wind, and oceanic currents [39]. For instance, plastic pellets that travelled across the Indian Ocean coastlines from Indonesia and Malaysia to Somalia were found to have transboundary repercussions in the MV X-Press Pearl catastrophe [41]. Seafloor plastics are a common phenomenon in almost all major oceanic systems and depend on seasonality, local and global currents, bathymetry, and local sources. They are classified into three groups in the marine environment: (i) floating debris/garbage patches (aggregate with gyres, such as the Great Pacific garbage patch); (ii) shoreline plastics (plastic wastes on the coast, depending on circulation and seasonal weathering patterns, coastal tourism, and socioeconomic backgrounds such as urbanized areas and changes in social habits) [42].

1.6   Plastics in the Atmosphere

Both manmade and natural polymers make up airborne microplastics. The bulk of airborne microplastics are around 20 μm in size, although they can range from 20 to 500 μm. Airborne microplastics come in a variety of colours (black and white) and shapes (fibre, the most common), including film, fragment, foam, granule, sphere, and fibre. One of the main sources of airborne microplastics is synthetic textiles, particularly the tiny fibres used in apparel [43]. For instance, around 1100 airborne microplastics are produced by 1 g of acrylic fabric. Activities such as drying or putting on clothes can release these fibres into the air, particularly from soft furnishings like curtains and carpets. Other sources of airborne microplastics include industrial emissions and the breakdown of bigger plastic objects. Generally speaking, indoor airborne microplastic concentrations are significantly higher than outside concentrations [44]. Rainfall, wind, pollution concentration, humidity, local circumstances, and particle size are some of the elements that affect the dispersion and subsequent deposition of airborne particles in atmospheric fallout. Microplastics are lifted from the ground into the air by the wind, which makes it easier for them to travel to far-off places. These airborne plastics contaminate terrestrial and aquatic environments when they are deposited. Thus, this dynamic cycle contributes to the presence of plastics in the total environmental landscape by exchanging them among the air, terrestrial, and aquatic habitats.

1.7   Biological Impacts

By limiting their growth, capacity for reproduction, and long-term toxicity, microplastics can have biological effects on living things [45]. Polyvinyl chloride (PVC)-derived microplastics disrupt oviposition, growth, oxidative damage, and biological enzyme activity [46]. According to [47], lugworms' gut tissues containing microplastics and additives have biological effects. Furthermore, there are negative interactions between microplastics and insects such as Culex mosquitoes [37]. In mice, microplastics also result in oxidative stress and problems with lipid and energy metabolism. Numerous marine species are impacted by plastic waste worldwide, including roughly 43% of marine mammals, 44% of seabirds, 86% of sea turtles, and a variety of fish and crustacean species, according to published research [48]. Ingestion, entanglement, reduced locomotion, habitat loss and feeding, low reproduction, ulcer, laceration, and mortality are the main biological effects of plastic debris in marine and other aquatic ecosystems [49]. For instance, consuming results in poor stomach capacity, internal injuries and blockages, and stunted growth. Strangulation, decreased feeding efficiency, and, in rare instances, drowning, are the outcomes of entanglement. Because sea turtles' major food source is gelatinous prey, or jellyfish, they mistakenly believe that plastic bags and other floating plastic trash are part of their diet. By considering them as food, seabirds consume a variety of plastic detritus, such as balloons, plastic pellets, and hard plastic fragments. Accordingly, the primary issues with seabirds are low body weight, impaired fat deposition, decreased capacity for reproduction, low fitness, and harm to the digestive system. Microplastics mimic phytoplankton, and occasionally plastics contaminate fish and cetaceans' preferred food sources [50]. The impact of plastic waste on cetaceans can result in direct mortality or weakening and increased susceptibility to disease or predators [51]. According to a study conducted in the North Pacific Central Gyre, 35% of the 670 fish in the study had 1375 plastic fragments in their bodies overall, or roughly 2.5 pieces per fish. The majority of those plastics are blue, white, and transparent, colours found in plankton, which is the main food source for fish [51]. As a result, plastic pieces may accumulate in food chain predators [52]. Specifically, polystyrene (PS)-derived nanoplastics affect the cellular and organizational levels and reduce the filter-feeding rate of blue mussels. Additionally, polystyrene microplastics decrease the size and quantity of oyster egg cells, sperm motility, and larval population. They also slow down Gammarus pulex growth [53]. Additionally, micro- and nanoplastics alter the sex ratio of turtle eggs during incubation and have an impact on biological processes connected to temperature. Plastics interact with both land and aquatic flora in addition to wildlife. According to [37], preliminary laboratory results indicate that plastic residues in the soil may have a detrimental effect on aquatic plants by shortening their shoots and roots. When animals encounter loops or gaps in the debris, they become entangled and may suffocate, drown, strangle, become unable to flee from predators, have reduced mobility, be unable to migrate, or even die. Fishing nets and other plastic items, such bands, collars, and straps, that are discarded by marine activities, may have an impact on the progeny of different marine animals. As the animals get bigger, these objects could snag on their necks or bodies, causing them to become constricted and potentially strangling [48]. Ingestion and individual organisms are tightly related. Significantly, entanglement and ingestion have a detrimental effect on fulmar, seals, sea lions, puffin, albatross, right whales, green sea turtles, hawksbill turtles, and greater shearwater. Physical and physiological consequences are two interrelated ways that consuming marine trash affects the health of wildlife [54]. The physical repercussions include (i) cuts and lesions-which happen when sharp objects pierce the intestinal lining and cause ulcers, lesions, inflammation, and infection. (ii) Blockage-when indigestible sheets and plastic bags become stuck in the stomach, exposing organs to digestive fluids and creating a false impression of fullness; and (iii) retention-when debris remains in the digestive tract for an extended period of time. Eventually, all of these physical effects lead to physiological repercussions, such as effects on nutrition, development, immunity, and toxicity. Furthermore, the chemical toxicants linked to plastic have an impact on population dynamics, reproductive cycles, and animal development, which may have long-term effects [55].

References

1. Torres-Agullo, A., Karanasiou, A., Moreno, T., Lacorte, S., 2021. Overview on the occurrence of microplastics in air and implications from the use of face masks during the COVID-19 pandemic. Sci. Total Environ. 800, 149555 https://doi.org/10.1016/j.scitotenv.2021.149555.
2. Hassan, T., Srivastwa, A.K., Sarkar, S., Majumdar, G., 2022. Characterization of plastics and polymers: A comprehensive study. IOP Conf. Series: Mater. Sci. Eng. 1225, 012033 https://doi.org/10.1088/1757-899x/1225/1/012033.
3. Khoaele, K.K., Gbadeyan, O.J., Chunilall, V., Sithole, B., 2023. The devastation of waste plastic on the environment and remediation processes: A critical review. Sustainability 15, 5233. https://doi.org/10.3390/su15065233.
4. Shah, M., Rajhans, S., Himanshu, P.A., Archana, U.M., 2021. Bioplastic for future: A review then and now. World J. Adv. Res. Rev. 9, 56–67. https://doi.org/10.30574/wjarr.2021.9.2.0054.
5. Narancic, T., Cerrone, F., Beagan, N., O'Connor, K.E., 2020. Recent advances in bioplastics: Application and biodegradation. Polymers 12, 920. https://doi.org/10.3390/polym12040920.
6. Pan, D., Su, F., Liu, C., Guo, Z., 2020. Research progress for plastic waste management and manufacture of value-added products. Adv. Compos. Hybrid Mater. 3, 443–461. https://doi.org/10.1007/s42114-020-00190-0.
7. Ilyas, M., Ahmad, W., Khan, H., Yousaf, S., Khan, K., Nazir, S., 2018. Plastic waste as a significant threat to environment – a systematic literature review. Rev. Environ. Health 33, 383–406. https://doi.org/10.1515/reveh-2017-0035.
8. Chandran, M., Tamilkolundu, S., Murugesan, C., 2020. Conversion of plastic waste to fuel. Plastic Waste Recycl. 14, 385–399. https://doi.org/10.1016/B978-0-12-817880-5.00014-1.
9. Panda, A.K., Singh, R.K., Mishra, D.K., 2010. Thermolysis of waste plastics to liquid fuel: a suitable method for plastic waste management and manufacture of value added products– a world prospective. Renew. Sustain. Energy Rev. 14, 233–248. https://doi.org/10.1016/j.rser.2009.07.005.

10. Lebreton, L., Andrady, A., 2019. Future scenarios of global plastic waste generation and disposal. Palgrave Commun. 5, 6. https://doi.org/10.1057/s41599-018-0212-7.

11. Worgull, M.; Kolew, A.; Heilig, M.; Schneider, M.; Dinglreiter, H.; Rapp, B. Hot Embossing of High Performance Polymers. Microsyst. Technol. 2011, 17(4), 585–592. https://doi.org/10.1007/s00542-010-1155-0.

12. Peng, L.; Wu, H.; Shu, Y.; Yi, P.; Deng, Y.; Lai, X. Roll-to-Roll Hot Embossing System with Shape Preserving Mechanism for the Large-Area Fabrication of Microstructures. Rev. Sci. Instrum. 2016, 87(10), 105120. https://doi.org/10.1063/1.4963907.

13. Aklonis, J. J.;. Time-Temperature Correspondence. In Introduction to Polymer Viscoelasticity, 3rd ed.; Shaw, M. T., MacKnight, W. J., Eds.; John Wiley & Sons: Hoboken, NJ, USA, 2005; pp 107–128.

14. Brahney, J., Hallerud, M., Heim, E., Hahnenberger, M., & Sukumaran, S. (2020). Plastic rain in protected areas of the United States. Science, 368(6496), 1257–1260. https://doi.org/10.1126/science.aaz5819.

15. Peng, X., Chen, M., Chen, S., Dasgupta, S., Xu, H., Ta, K., & Bai, S. (2018). Microplastics contaminate the deepest part of the world's ocean. Geochemical Perspectives Letters, 9 (1), 1–5. https://doi.org/10.7185/geochemlet.1829.

16. Leslie, H., et al. (2022). Discovery and quantification of plastic particle pollution in human blood. Environment International. https://doi.org/10.1016/j.envint.2022.107199.

17. Wilcox, C., Van Sebille, E., & Hardesty, B. D. (2015). Threat of plastic pollution to seabirds is global, pervasive, and increasing. Proceedings of the National Academy of Sciences, 112(38), 11899–11904.

18. Bucci, K., Tulio, M., & Rochman, C. M. (2020). What is known and unknown about the effects of plastic pollution: A meta-analysis and systematic review. Ecological Applications, 30(2), e02044.

19. Elliff, C. I., Mansor, M. T. C., Feodrippe, R., & Turra, A. (2021). Microplastics and the UN Sustainable Development Goals. In T. Rocha-Santos, M. Costa, & C. Mouneyrac (Eds.), Handbook of Microplastics in the Environment. Cham: Springer. https://doi.org/10.1007/978-3-030-10618-8_24-1.

20. Villarrubia-Gomez, P., Cornell, S. E., & Fabres, J. (2018). Marine plastic pollution as a planetary boundary threat: The drifting piece in the sustainability puzzle. Marine Policy, 96, 213–220. https://doi.org/10.1016/j.marpol.2017.11.035.

21. Amaral-Zettler, L. A., Zettler, E. R., & Mincer, T. J. (2020). Ecology of the plastisphere. Nature Reviews Microbiology, 18, 139–151. https://doi.org/10.1038/s41579-019-0308-0.

22. Haward, M. (2018). Plastic pollution of the world's seas and oceans as a contemporary challenge in ocean governance. Nature Communication, 9, 667. https://doi.org/10.1038/s41467-018-03104-3.

23. Lohr, A., Savelli, H., Beunen, R., Kalz, M., Ragas, A., & Van Belleghem, F. (2017). Solutions for global marine litter pollution. Current opinion in environmental sustainability, 28, 90–99. https://doi.org/10.1016/j.cosust.2017.08.009.

24. Jambeck, J., et al. (2015). Plastic waste inputs from land into the ocean. Science, 347 (6223), 768–771. https://doi.org/10.1126/science.1260352.

25. Stoett, P. & Vince, J. (2019). The Plastic-Climate Nexus: Linking Science, Policy, and Justice. In P. Harris (ed.), Climate Change and Ocean Governance: Politics and Policy for Threatened Seas, pp. 345–361. Cambridge University Press, 2019.

26. McIntyre, O. (2020). Addressing marine plastic pollution as a 'wicked' problem of transnational environmental governance. Environmental Liability: Law, Policy and Practice, 25(6), 282–295.

27. Stoett, P., & Omrow, D. (2021). Spheres of Transnational Ecoviolence: Environmental Crime, Human Security, and Justice. New York: Palgrave MacMillan/Springer Nature.

28. Fox, O., & Stoett, P. (2016). Citizen participation in the un sustainable development goals consultation process: toward global democratic governance? Global Governance: A Review of Multilateralism and International Organizations, 22(4), 555–574. https://doi.org/10.1163/194 26720-02204007.

29. Nilsson, M., Chisholm, E., Griggs, D., Howden-Chapman, P., McCollum, D., Messerli, P., Neumann, B., Stevance, A. S., Visbeck, M., & Stafford-Smith, M. (2018). Mapping interactions between the sustainable development goals: Lessons learned and ways forward. Sustainability Science, 13, 1489–1503. https://doi.org/10.1007/s11625-018-0604-z.

30. Walker, T. (2021). (Micro)plastics and the UN sustainable development goals. Current Opinion in Green and Sustainable Chemistry, 30, Article 100497. https://doi.org/10.1016/j.cogsc.2021. 100497.

31. Zamboni, et al. (2021). Unfolding differences in the distribution of coastal marine ecosystem services values among developed and developing countries. Ecological Economics, 189, Article 107151. https://doi.org/10.1016/j.ecolecon.2021.107151.

32. United Nations (2017). Resolution adopted by the General Assembly on 6 July 2017- 71/ 313. Work of the Statistical Commission pertaining to the 2030 Agenda for Sustainable Development. https://undocs.org/A/RES/71/313. Accessed 9 Apr 2020.

33. United Nations (2022). The Sustainable Development Goals Report 2022. Available at: https:// unstats.un.org/sdgs/report/2022/ Accessed 9 November 2022.

34. Geyer, R., Jambeck, J.R., Law, K.L., 2017. Production, use, and fate of all plastics ever made. Sci. Adv. 3 https://doi.org/10.1126/sciadv.1700782.

35. Huang, S., Wang, H., Ahmad, W., Ahmad, A., Vatin, N.I., Mohamed, A.M., Deifalla, A.F., Mehmood, I., 2022. Plastic waste management strategies and their environmental aspects: a scientometric analysis and comprehensive review. Int. J. Environ. Res. Public Health 19, 4556. https://doi.org/10.3390/ijerph19084556.

36. Gabbott, S., Key, S., Russell, C., Yonan, Y., Zalasiewicz, J., 2020. The geography and geology of plastics. Plastic Waste Recycl. 3, 33–63. https://doi.org/10.1016/B978-0-12-817880-5.00003-7.

37. Hurley, R., Horton, A., Lusher, A., Nizzetto, L., 2020. Plastic waste in the terrestrial environment. Plastic Waste Recycl. 7, 163–193. https://doi.org/10.1016/B978-0-12-817880-5.000 07-4.

38. Azevedo-Santos, V.M., Brito, M.F.G., Manoel, P.S., Perroca, J.F., Rodrigues-Filho, J.L., Paschoal, L.R.P., Gonçalves, G.R.L., Wolf, M.R., Blettler, M.C.M., Andrade, M.C., Nobile, A.B., Lima, F.P., Ruocco, A.M.C., Silva, C.V., Perbiche-Neves, G., Portinho, J. L., Giarrizzo, T., Arcifa, M.S., Pelicice, F.M., 2021. Plastic pollution: a focus on freshwater biodiversity. Ambio 50, 1313–1324. https://doi.org/10.1007/s13280-020-01496-5.

39. Eriksen, M., Lebreton, L.C.M., Carson, H.S., Thiel, M., Moore, C.J., Borerro, J.C., Galgani, F., Ryan, P.G., Reisser, J., 2014. Plastic pollution in the world's oceans: more than 5 trillion plastic pieces weighing over 250,000 tons float at sea. Public Libr. Sci. One 9. https://doi.org/10.1371/ journal.pone.0111913.

40. Cózar, A., Echevarría, F., González-Gordillo, J.I., Irigoien, X., Úbeda, B., Hernandez-León, S., Palma, A.T., Navarro, S., García-de-Lomas, J., Ruiz, A., Fernandez-de-Puelles, M.L., Duarte, C.M., 2014. Plastic debris in the open ocean. Proc. Natl. Acad. Sci. U. S. A. 111, 10239–10244. https://doi.org/10.1073/pnas.131470511.

41. Perera, U.L.H.P., Subasinghe, H.C.S., Ratnayake, A.S., Weerasingha, W.A.D.B., Wijewardhana, T.D.U., 2022. Maritime pollution in the Indian Ocean after the MV XPress Pearl accident. Mar. Pollut. Bull. 182, 114301 https://doi.org/10.1016/j.marpolbul.2022.114301.

42. Welden, N.A., 2020. The environmental impacts of plastic pollution. Plastic Waste Recycl. 8, 195–222. https://doi.org/10.1016/B978-0-12-817880-5.00008-6.

43. Huang, Y., Qing, X., Wang, W., Han, G., Wang, J., 2020. Mini-review on current studies of air-borne microplastics: Analytical methods, occurrence, sources, fate and potential risk to human beings. TrAC Trends Anal. Chem. 125, 115821 https://doi.org/10.1016/j.trac.2020.115821.

44. Wang, X., Li, P., Su, M., Zou, X., Duan, L., Zhang, H., 2021. Characteristics of plastic pollution in the environment: a review. Bull. Environ. Contam. Toxicol. 107, 577–584. https://doi.org/10.1007/s00128-020-02820-1.

45. Zhang, W., Sik, O.Y., Bank, M.S., Sonne, C., 2023. Macro- and microplastics as complex threats to coral reef ecosystems. Environ. Int. 174, 107914 10.1016/ j.envint.2023.107914.

46. Espinosa, C., Cuesta, A., Esteban, M.A., ´2017. Effects of dietary polyvinylchloride microparti-cles on general health, immune status and expression of several genes related to stress in gilthead seabream (Sparus aurata L.). Fish Shellfish Immunol. 68, 251–259. https://doi.org/10.1016/j.fsi.2017.07.006.

47. Browne, M.A., Niven, S.J., Galloway, T.S., Rowland, S.J., Thompson, R.C., 2013. Microplastic moves pollutants and additives to worms, reducing functions linked to health and biodiversity. Curr. Biol. 23, 2388–2392. https://doi.org/10.1016/j.cub.2013.10.012.

48. Thushari, G.G.N., Senevirathna, J.D.M., 2020. Plastic pollution in the marine environment. Heliyon 6, E04709. https://doi.org/10.1016/j.heliyon.2020.e04709.

49. Schnurr, R.E.J., Alboiu, V., Chaudhary, M., Corbett, R.A., Quanz, M.E., Sankar, K., Srain, H.S., Thavarajah, V., Xanthos, D., Walker, T.R., 2018. Reducing marine pollution from single-use plastics (SUPs): A review. J. Mar. Pollut. Bull. 137, 157–171. https://doi.org/10.1016/j.marpol bul.2018.10.001.

50. Boerger, C.M., Lattin, G.L., Moore, S.L., Moore, C.J., 2010. Plastic ingestion by planktivorous fishes in the North Pacific Central Gyre. Mar. Pollut. Bull. 60, 2275–2278. https://doi.org/10.1016/j.marpolbul.2010.08.007.

51. Sigler, M., 2014. The effects of plastic pollution on aquatic wildlife: Current situations and future solutions. Water Air Soil Pollut. 225, 2184. https://doi.org/10.1007/s11270-014-2184-6.

52. Teuten, E.L., Saquing, J.M., Knappe, D.R.U., Barlaz, M.A., Jonsson, S., Bjorn, ¨A., Rowland, S.J., Thompson, R.C., Galloway, T.S., Yamashita, R., Ochi, D., Watanuki, Y., Moore, C., Viet, P.H., Tana, T.S., Prudente, M., Boonyatumanond, R., Zakaria, M.P., Akkhavong, K., Ogata, Y., Hirai, H., Iwasa, S., Mizukawa, K., Hagino, Y., Imamura, A., Saha, M., Takada, H., 2009. Trans-port and release of chemicals from plastics to the environment and to wildlife. Philos. Trans. R. Soc. London 364, 2027–2045. https://doi.org/10.1098/rstb.2008.0284.

53. Redondo-Hasselerharm, P.E., Falahudin, D., Peeters, E.T.H.M., Koelmans, A.A., 2018. Microplastic effect thresholds for freshwater benthic macroinvertebrates. Environ. Sci. Tech. 52, 2278–2286. https://doi.org/10.1021/acs.est.7b05367.

54. Jacob, H., Besson, M., Swarzenski, P.W., Lecchini, D., Metian, M., 2020. Effects of virgin micro- and nanoplastics on fish: Trends, meta-analysis, and perspectives. Environ. Sci. Tech. 54, 4733–4745. https://doi.org/10.1021/acs.est.9b05995.

55. Degli, E.M., Morselli, D., Fava, F., Bertin, L., Cavani, F., Viaggi, D., Fabbri, P., 2021. The role of biotechnology in the transition from plastics to bioplastics: An opportunity to reconnect global growth with sustainability. FEBS Open Bio 11, 967–983. https://doi.org/10.1002/2211-5463.13119.

Microbial Resources

2.1 Microbes and Their Role in Plastic Degradation

Microorganisms play a significant role in the biodegradation of plastics, including ambient bacteria (in vitro) and gut microbes of insects (in vivo). Microbial deterioration under ecological circumstances in vitro is prolonged for primary plastics at degradation rates depending on a month or even a year [1, 2]. However, recent discoveries indicate that the fast biodegradation of specific plastics, such as PS, PE, and PUR, in some invertebrates, mainly insects, could be enhanced at hourly rates; biodegradation in insects is likely to be gut microbial-dependent or synergetic bio reactions in animal digestive systems. Mechanical, photochemical, thermal, and metabolic methods can all be used to destroy plastics in the natural environment. Photochemistry is the most effective chemical degradation mechanism for plastics in nature [3, 4] (Fig. 2.1).

Photochemistry is the most efficient natural chemical degradation process for plastics. Thermal oxidation occurs slowly at ambient temperature [5, 6] but when temperature rises, the efficiency of thermal oxidation increases as well. Using a sophisticated enzyme system, microorganisms may effectively break down and extract energy from plastic polymers [2]. Although the microbial decomposition rate is slow, it is considered a more environmentally acceptable method of removing plastic trash [7]. It is unclear whether the energy derived from plastic decomposition can support microbiological activity, particularly growth.

In 1975, Flavobacterium was reported to degrade nylon in wastewater pools from a nylon industry. Later, an increasing number of microorganisms were discovered to degrade polymers found in natural environments such as soil, seawater, sludge, and compost [8]. The narrative of insects and plastics begins with customers complaining that insects eat chocolate-based packaging [9]. Insect degradation capacity was assessed based on

© The Author(s), under exclusive license to Springer Nature Switzerland AG 2025
J. A. Parray and W.-J. Li, *Microbial and Enzyme-Based Technology for Plastic Biodegradation*, Synthesis Lectures on Chemical Engineering and Biochemical Engineering, https://doi.org/10.1007/978-3-031-84437-9_2

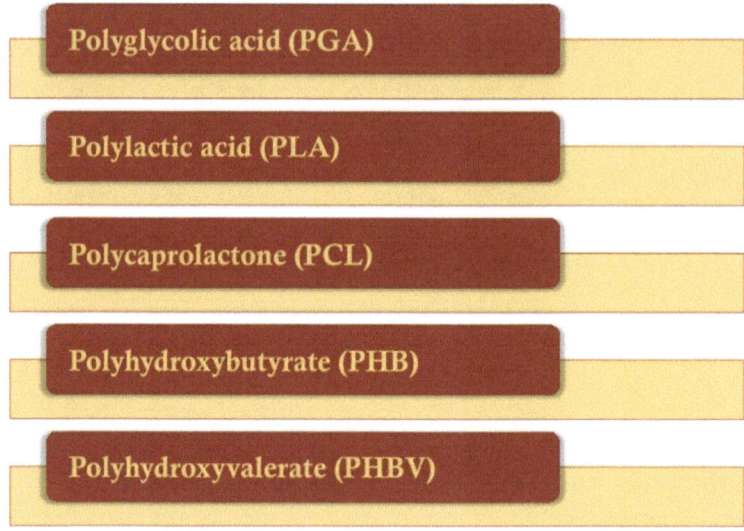

Fig. 2.1 Common forms of biodegradable plastics

observations of insects destroying and devouring plastic packaging materials. Today, scientists have screened various environmental microorganisms (in vitro) and gut microbes of insects (in vivo) to digest plastics [9]. Plastic degradation happens via various methods, including photooxidation, and thermal-, catalytic-, and biodegradation, with the latter creating CO_2 and water while providing environmental advantages [9]. Weathering and biological activities have the potential to degrade ecological plastics into microplastics (MPs) and nanoplastics (NPs). Although plastic garbage appears nonexistent, it degrades into MPs and NPs, which poses health dangers. Some creatures can break plastics, including Antarctic krill, which convert MPs into NPs through digestion, creating toxicological concerns. Many bacteria and fungi can degrade MPs in laboratory and natural environments [10].

Microbes exist everywhere in the biosphere, impacting the environment in which they flourish. Microbes can have both positive and negative effects on their surroundings. The most crucial job of microorganisms on Earth is their ability to degrade organic materials and recycle the fundamental elements (carbon (C), oxygen (O), and nitrogen (N)) that comprise all living systems. Plastic biodegradation is the breakdown of complex polymers into simpler monomers, determined by various parameters such as shape, substrate availability, polymer molecular weight, and surface features [11].

2.2   Microbes and Their Potential Application

Various abiotic and biotic variables contribute significantly to the biodegradation of recalcitrant PE in the environment [12]. Biodegradation studies have been conducted using either pure cultures capable of degrading PE or complex microbial communities from various terrestrial (landfill soil, composting) and marine habitats. Furthermore, customized microbial consortia could aid in the biodegradation of PE (Fig. 2.2) [13].

2.2.1   Bacteria

Bacteria are crucial and abundant organisms in nature that can degrade plastic. Recently, microorganisms deleting plastic have been isolated from marine, trash, soil, and compost habitats. Notable isolates include *Azotobacter sp., Bacillus megaterium, Ralstonia eutropha,* and several *Pseudomonas* and *Halomonas species. Bacillus brevis, thermophilic bacteria,* can degrade polylactic acid. Furthermore, research indicates that *Arthrobacter sp., Acinetobacter baumannii,* and others can break down polyethene [12–14].

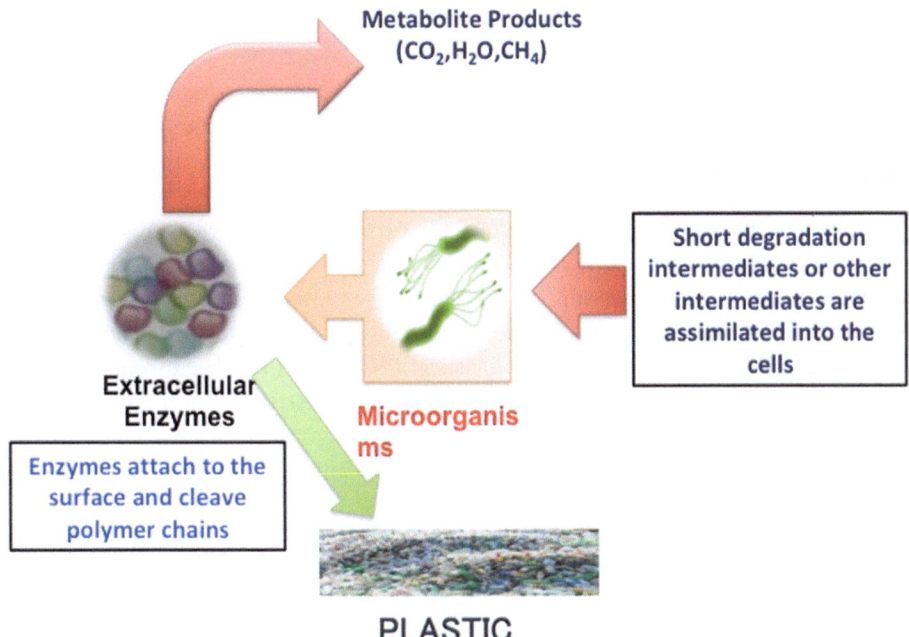

Metabolite Products
(CO$_2$,H$_2$O,CH$_4$)

Short degradation intermediates or other intermediates are assimilated into the cells

Extracellular Enzymes

Microorganisms

Enzymes attach to the surface and cleave polymer chains

PLASTIC

Fig. 2.2 Microbial degradation of plastics in natural environment

Over 20 bacterial taxa have been found to break down various kinds of PE. These include a variety of Gram-negative and Gram-positive species from the genera *Pseudomonas, Ralstonia, Stenotrophomonas, Klebsiella, Acinetobacter, Rhodococcus, Staphylococcus, Streptococcus, Streptomyces, Bacillus,* and so on [12–16]. Most bacterial strains can degrade the surface and establish a biofilm on PE. Studies on the various activities of the Pseudomonas genus have been conducted to study their ability to digest and metabolize a variety of synthetic plastic polymers and by products. *Pseudomonas* species can degrade and metabolize polymers using extracellular oxidative and hydrolytic enzymes, promoting polymer fragment uptake and degradation while modulating the interface between biofilms and polymer surfaces [17]. *Pseudomonas fluorescens* completely degraded polyethene (PE) in water, particularly with surfactants. This highlights their role in polymer biodegradation. Tribedi and Sil [18] discovered that mineral oil improved biofilm growth on low-density polyethene (LDPE) when treated with *Pseudomonas* sp. strain AKS2, resulting in a 5% degradation over 45 days, but Tween 80 inhibited this process. *Brevibacillus borstelensis*, a thermophilic bacterium, lowered the molecular weight of biodegradable low-density polyethylene (BLDPE) by 30% within 30 days. Additionally, *Rhodococcus ruber* (C208), a biofilm producer, destroyed PE at 0.86% weekly, most likely due to its hydrophobic cell surface. After thermal treatment, Awasthi et al. [19] found that *Klebsiella pneumoniae* damaged HDPE. This strain could securely cling to HDPE surfaces, increasing biofilm thickness while lowering the weight and tensile strength of the HDPE film by 18.4 and 60%, respectively, in 60 days. SEM and atomic force microscopy (AFM) images of subsurface corrosion, fractures, and surface roughness caused by bacteria show that an HDPE film may be biodegrading.

Pseudomonas Species

Pseudomonas spp. currently accounts for 21% of the bacterial genera related to plastic degradation [20]. The first microbial experiments on plastic biodegradation were also conducted using *Pseudomonas* spp. [21]. Plastics have been degraded using several Pseudomonas species. Among them, Pseudomonas aeruginosa has attracted much attention and has been shown to damage various plastics. *P. aeruginosa* isolated from superworm intestines caused daily weight loss of 0.64, 0.098, and 0.025% for PE, PS, and PP, respectively [22]. *P. aeruginosa* isolated from trash dumps degraded polyethene (PE) at rates of 6.5 and 8.7% after 60 days of incubation in minimal salt medium (MSM) and Bushnell-Haas broth (BHM), respectively [23].

Pseudomonas spp. has been commonly found in marine plastic biofilms. However, there have been fewer instances of their strains being isolated individually to break down plastic than from terrestrial sources [24]. *Pseudomonas aestusnigri,* isolated from sea sand samples, can successfully degrade PET using a new carboxylic acid ester hydrolase [25]. *Pseudomonas rhodesiae* isolated from Brazilian deep-sea sediments has been shown to develop biofilms on high-density polyethene (HDPE), causing structural alterations in the plastic [25]. Microbial consortia formed by *P. putida* and *P. stutzeri* isolated from

Bangalore Lake could degrade low-density polyethene (LDPE), demonstrating greater degradability and 90% weight loss after 40 days of incubation [26].

Bacillus sps

Bacillus spp. can effectively decompose several types of plastics, including *Bacillus cereus, Bacillus safensis,* and *Bacillus subtilis.* For example, *B. cereus* isolated from mangroves in Peninsular Malaysia caused weight losses of 1.6, 6.6, and 7.4% for PE, PET, and PS in 40 days [27]. *B. subtilis* H1584 isolated from pelagic waters degraded LDPE, causing a weight loss of 1.75% in 30 days. After 90 days of incubation, *B. subtilis* isolated from saltwater at depths ranging from 0 to 30 cm lost 1.54% of its LDPE weight. *Bacillus paralicheniformis* G1, isolated from deep-sea sediment, was reported to degrade PS by about 18% in the first 30 days and 34% after 60 days. Plastic-degrading Bacillus isolated from terrestrial habitats is significantly more diversified; for example, *B. cereus* isolated from landfills showed a 1.78% breakdown rate of HDPE in 30 days, whereas B. cereus isolated from cow manure produced a weight loss of 5.9% HDPE in 83 days. *B. safensis* from landfill soil reduced the weight of PLA 1006 by 8% after 30 days. After 60 days of incubation, two strains of *B. amyloliquefaciens* (BSM-1) and *B. amyloliquefaciens* (BSM-2) isolated from municipal solid soil degraded LDPE by 11% and 16%, respectively [27, 28].

Alcanivorax sps

The genus Alcanivorax has an extensive distribution in the ocean and is well-known for hydrocarbon degradation. Their presence in biofilms of marine plastics indicated their ability to break down plastics. *Alcanivorax sp.* 24, isolated from marine plastic trash, reduced mass by 0.9% following 34 days of incubation with LDPE [29]. *Alcanivorax borkumensis,* isolated from the Mediterranean Sea, may develop biofilms on LDPE, resulting in a 3.5% weight loss after 80 days. *Alcanivorax wenustensis,* a deep-sea organism, has been shown to break down polycaprolactone (PCL) and roughen its surface. *Alcanivorax xenomutans* isolated from mangroves were found to effectively break down PS based on strain growth [30, 31].

Actinomycetes

Plastic-degrading actinomycetes are primarily isolated from soil. However, several actinomycetes from coastal habitats have been isolated and shown to destroy plastics well. *Kocuria palustris,* isolated from Arabian Sea seawater, could lower PE weight by 1% after 75 days of incubation. *Rhodococcus ruber* isolated from mangrove sediment degraded PP with a 6.4% weight loss after 40 days of incubation. *Rhodococcus pyridinivorans* P23, isolated from deep-sea sediment, could reduce PET's weight by 4.28% after five weeks of incubation. *Nocardioides marinus,* isolated from Pacific Ocean deep-sea sediments, could lower PET weight by 1.2–1.3% in 30 days [32, 33].

Gordonia Sinensis, Gordonia mangrove, and *Gordonia bronchialis,* all isolated from mangroves, showed PS degradation rates ranging from 4.69 to 7.73% in one month. *Streptomyces gougerotti, Micromonospora matsumotoense,* and *Nocardiopsis prasina* isolated from the water were able to degrade LDPE, PS, and polylactic acid (PLA) to varied degrees. Actinomycetes are known to break down polylactic acid (PLA). A strain of Amycolatopsis sp. was shown to eliminate 60% of the PLA film (100 mg) in liquid culture at 30 °C after 14 days. *Amycolatopsis sp., Amycolatopsis* sp., *Saccharothrix, Kibdelosporangium aridum, Actinomadura keratinilytica, Amycolatopsis thailandensis, Streptomyces bangladeshensis,* and *Streptomyces thermoviolaceus subsp. thermoviolaceus* are among the actinomycetes known to degrade Polybutylene succinate (PBS), PCL, and PLA [34–36].

Other Bacterial Species
Several *Proteobacteria, Firmicutes,* and *Actinobacteria* species can biodegrade plastics—Lysinibacillus sp. JJY0216 from soil degraded 9 and 4% of PE and PP, respectively, after 26 days. In one month, *Achromobacter denitrificans* Eb113 degraded LDPE and PVC by 6.5 and 22.3%, respectively. *Alcaligenes faecalis* destroyed LLDPE by 3.5%, HDPE by 5.8%, and polyester by 17.3% after 40 days [37]. *Meyerozyma guilliermondii* and *Serratia marcescens* from wax worm guts resulted in PE losses of 13.9 and 3.57%, respectively, after 60 days. After 35 days, *Brevibacillus brevis* from soil reduced the weight of nylon by 6.6 by 22%. Acinetobacter from *Tribolium castaneum* larvae resulted in a 12.14% decrease in PS at 60 days. *Marinobacter gudaonensis, Thalassospira xiamenensis,* and *Marinobacter sediminum,* derived from deep-sea sediments, reduced PET weight by 1.2–1.3% within 30 days [37, 38].

2.2.2   Fungi

Fungi are essential in environmental plastic degradation because they migrate across substrates using their filamentous network structure, investigating and growing in areas other microbes find difficult to reach [39]. Previously, mainly terrestrial fungi have documented biodegradation by fungi with polyesterase activities. Several marine plastic-degrading fungi have been identified. For example, the marine red yeast *Rhodotorula mucilaginosa,* isolated from the North Sea plastic trash, can successfully break down PE using an isotopically labelled technique [40]—*Penicillium* spp. isolated from the Red Sea Coast can bind to LDPE films and develop rapidly. *Alternaria alternata* FB1, a marine fungus isolated from plastic waste, can degrade PE by creating holes in the film [41].

With a mass change of 56.7% in 14 days, *Zalerion maritimum,* isolated from Portuguese coastal waters, had exceptional activity in breaking down PE microplastics. By releasing extracellular materials, the marine fungus *Cladosporium halotolerans* 6UPA1, isolated from deep-sea sediments, colonizes PU foam and breaks down plastic [42]. As evidenced by weight loss and morphological changes like cracks and fissures, several terrestrial *Aspergillus* species, including *Aspergillus nomius* RH06 and *Aspergillus clavatus* JASK1, can break down LDPE polymers. Significant plastic degradability is also exhibited by other terrestrial fungal species, such as *Trichoderma* sp., *Monascus* sp., *Clitocybe* sp., and *Penicillium* spp. and *Phanerochaete* sp. Fungal strains that have been linked to plastic degradation recently include *Aspergillus versicolor, A. flavus, Chaetomium sp.,* and *Mucor circinelloides sp. A. niger, A. cremeus, A. nidulans, A. ornatus, A. candidus, A. glaucus,* and amongst, *A. flavus are the most common fungal species found to degrade polythene bags.* Besides Numerous researches have been reported regarding the role of Fungal species in plastic degardation like wise polylactic acid (PLA) is also degraded by *Penicillium roqueforti* and *Tritirachium album* [42–46].

The PVC, polyesters (PS) and polyhydroxybutyrates (PHBs) are subjected to degradation by various fungal genera like *Acremonium, Mucor, Cladosporium, Debaryomyces, Paecilomyces, Fusarium, Pullularia, Emericellopsis, Eupenicillium, Penicillium, Verticillium,* and *Rhodosporidium. Like wise Aureobasidium, Aspergillus, Cryptococcus, Chaetomium, Rhizopus, Fusarium, Thermoascus,* and *Penicillium* degrade Polycaprolactone (PCL) and Polyester acetals (PEA) [42]. Fungi like *Aspergillus terreus, Aspergillus flavus, Alternaria solani, Aspergillus fumigates,* and *Spicaria* sp. have been isolated from plastic-dumped soil [43]. Because they can adhere to the hydrophobic surface of the polymers producing extracellular enzymes that target insoluble fibres, and endure demanding growth circumstances, fungi are generally believed to be more effective than bacteria at degrading PE. Measurement of weight loss is a popular technique for examining PE biodegradation. For instance, it has been discovered that *A. niger* and *A. japonicas* biodegrade LDPE in a lab setting, reducing the dry weight by 5.8 and 11.1% monthly, respectively [46]. Das and Kumar investigated the microbial degradation of LDPE by *Aspergillus* and *Fusarium* sp. [32]. Of these, *Aspergillus* sp. FSM-3 and *Fusarium* sp. FSM-10 demonstrated a maximum weight loss of roughly 8–9%. In contrast, only 5% weight loss was. Using enrichment culture, Usha et al. [47] identified strains of *Aspergillus flavus* and *A. nidulans,* displaying the clearing zone surrounding their colonies on PE agar plates. PE-degrading fungi were isolated from mangrove soils. According to Yamada-Onodera et al. [48], *P. simplicissimum* degrades PE without additions. Esmaeili et al. [49] used a mineral medium containing PE powder as the only carbon source to extract *A. niger* from soils of landfills that contain PE trash. Several strains of *Chrysonilia, Aspergillus,* and *Penicillium* species have been isolated using a synthetic media based on SEM and AFM examination of the PE surface. *P. chrysogenum* NS10 (KU559907) and *P. oxalicum* NS4 (KU559906) are two of the isolated fungi whose HDPE and LDPE degradation has been assessed using the response surface methodology.

2.2.3   Algae

There have only been a few documented occurrences of algae causing plastic breakdown. Plastic is broken down by ligninolytic and exopolysaccharide enzymes produced by algae adsorbed on the surface [50]. By observing corrosion, abrasion, groove, and ridge patterns, it was discovered that the microalga *Uronema africanum*, isolated from waste plastic bags in a freshwater lake, degraded LDPE sheets in 30 days. Cyanobacteria were found on plastic biofilms during the examination, indicating they can break down plastic. There are fewer cyanobacteria known to break down plastics. For instance, it was discovered that *Oscillatoria subbrevis* and *Phormidium lucidum*, which were isolated from plastic waste in home wastewater, could break down LDPE because they colonized PE and used carbon without the need for pretreatment or pro-oxidant chemicals [51]. According to FTIR-ATR, SEM–EDX, and tensile strength, PET and PP can degrade after 112 days of contact with freshwater Spirulina isolates. There have been no reports of cyanobacteria isolated from marine sources degrading plastic. A marine diatom called PETaseR280A-FLAG demonstrated the capacity to break down PET. This is the only known instance of marine-sourced algae breaking down plastics, and it lays the groundwork for the eventual application of microalgae to address the plastic issue [51].

2.2.4   Enzymes

Enzymatic biocatalysis provides a sustainable substitute for plastic recycling and is essential to the biodegradation of polymers. In the biodegradation of plastics, enzymes break the chemical bond after first adsorption on the film surface with the surface-binding domain. The enzymes that break down PET have been studied the most out of all of them. After three weeks of incubation, hydrolases derived from the actinomycete *Thermobifida fusca* were shown to reduce weight by 50% in 2005 [52]. PET-degrading enzymes have now been identified, primarily consisting of PETase, MHETase, and carboxylic ester hydrolases (cutinases, carboxylesterase, and lipases). Cutinases and PETase have drawn the most interest among them. The breakdown activity of PET hydrolase PETase, identified from *Ideonella sakaiensis* in 2016, is at its peak at 40. The stability and activity of PET hydrolases have been improved using various techniques (chemical modification, PET pretreatment, protein engineering, etc.) to increase their capacity for depolymerization. Directed evolution was used to screen the best variation, DepoPETase, which demonstrated a 23.3 °C higher Tm value than the original PETase and could yield 1407-fold products towards amorphous PET film at 50 °C [52]. Modulating post-translation glycan modification increased the activity and thermostability of *I. sakaiensis* PETase, which can fully depolymerize untreated PET plastic in two to three days at 50 °C. According to a different study, two fusion proteins made by a leaf-branch compost cutinase and a carbohydrate-binding module might improve PET film breakdown efficiency by 3.7 and

24.2%, respectively [53]. The primary class of enzymes involved in the breakdown of PU is esterase. The PUR esterase isolated from *Comamonas acidovorans* TB-35 in 1998 could attach itself to the PUR surface and hydrolyze its ester linkages. This enzyme's ideal pH was 6.5, and its ideal temperature was 45 °C. The polyurethane was later found to be involved in the breakdown of PU after being isolated from Pseudomonas species, including *Pseudomonas fluorescens, Pseudomonas chlororaphis,* and *Pseudomonas* sp. AKS31 [53–55]. There aren't many studies on the characterization of particular enzymes for PE, PS, PP, and PVC. The oxidation and depolymerization of PE have been linked to peroxidase, laccase, manganese superoxide dismutase, and alkane hydroxylase also. For instance, after 24 h of incubation at 30 °C, the Antarctic marine *Psychrobacter* sp. NJ228 laccase can reduce the mass by 13.2%. Similarly, the PS breakdown process involves lipase, laccase, and oxidoreductases. Polystyrene breakdown by hydroquinone peroxidase derived from *Azotobacter beijerinckii* HM121 and C–C bonds was shown to be broken by alkane hydroxylases and monooxygenases, ring-hydroxylating dioxygenases. Both natural and artificial plastics are linked to or broken down by microorganisms such as actinomycetes, fungi, bacteria, and *saccharomonospora*. It has been discovered that nine fungal species and seventeen bacterial genera break down plastic [55–59] (Table 2.1).

Table 2.1 The degradation of polyethene (PE), low-density polyethene (LDPE), and high-density polyethene (HDPE) by organisms

Organism	Plastic-type	Degradation time (days)	Biodegradation efficiency (%)	References
Pseudomonas fluorescens	PE	270	18.0	[53]
Bacillus vallismortis bt-dsce01	LDPE	120	75.0	[54]
Klebsiella pneumoniae CH001	HDPE	60	18.4	[19]
Aspergillus oryzae strain A5; *Bacillus cereus* strain A5	LDPE	112; 112	36.4; 35.72	[55]
Trichoderma viride RH03; *Aspergillus nomius* RH06	LDPE	45; 45	5.13; 6.63	[56]
Bacillus sp. and *Paenibacillus* sp.	PE	60	14.7	[57]
Aspergillus flavus	HDPE	100	5.5	[58]
Bacillus siamensis	LDPE	90	8.46	[59]

2.3   Overview of Insect Gut Microbiome for Plastic Degradation

Numerous insect species, such as mealworms (*Tenebrio molitor* larvae), superworms (*Zophobas atratus* larvae), and greater wax moth larvae (*Galleria mellonella* L), have been shown to break down plastics. Notably, *Tenebrio molitor* shows exceptional plastic degrading efficiency. As per the recent study, mealworms can degrade polyester-PU foam with an efficiency of up to 67% [60].

In addition to degrading low-density polyethene (LDPE), *Tenebrio molitor* larvae also break down linear low-density polyethene (LLDPE) and high-density polyethene (HDPE). Molecular weight, branching number, and plastic-type affected depolymerization ability. According to Yang et al. [90] [60], yellow mealworms can fully break down PS into CO_2 and incorporate it into their biomass. According to reports, yellow mealworms are less capable of degrading PS than *Zophobas atratus* and *Tenebrio obscurus* [58, 59]. Insect plastic biodegradation is thought to be a more environmentally friendly alternative because *Zophobas atratus* larvae can break down PS and PE.

Acinetobacter has been identified from the larvae of *Tribolium castaneum*, capable of degrading PS. PE is consumed by *Plodia interpunctella*, but its stomach is degraded by *Enterobacter asburiae* and *Bacillus*. *Tribolium confusum* and *Achroia grisella* are two other invertebrates that aid in the biodegradation of plastic. Earthworms may accelerate biodegradation through soil microbial activity and bioturbation, and they have demonstrated the ability to accelerate the decomposition of biodegradable plastic. Although additional research is required, termites can break down plastics through their stomach microbiota [58–60].

Role of Waxworm and Gut Microbiome in Biodegradation of PE

Plodia interpunctella and *Galleria mellonella* larvae can break down LDPE without prior treatment. Because beeswax and PE are similar, these worms soften PE shopping bags and convert them to ethylene glycol, which helps with their metabolism. Waxworm studies have been criticized for lacking evidence of PE breakdown; nonetheless, surface damage observations and FTIR research suggest that PE breaks down by cleaving carbon–carbon bonds mechanically or enzymatically—PE biodegradation by *Enterobacter* sp. D1 from *G. mellonella* intestines. Two microorganisms that break down PE, *Bacillus* sp. YP1 and *Enterobacter asburiae* YT1, were isolated from the gut of *P. interpunctella* larvae in a different investigation. After 28 days of incubation, they produced surface breakdown and decreased the hydrophobicity of the PE film. These strains caused around $10.7 \pm 0.2\%$ and $6.1 \pm 0.3\%$ of the PE films (100 mg) to deteriorate after 60 days of incubation. These results illustrated the significance of moth larvae gut bacteria for PE biodegradation. Yang et al. [61] investigated the role of yellow mealworms, or *Tenebrio molitor* larvae, biodegraded PE and plastic mixes. Upon incubation with larvae, up to $49.0 \pm 1.4\%$ of the consumed PE was converted to CO_2. When mealworms were fed PE, their molecular weights decreased by $40.1 \pm 8.5\%$.

According to the gut microbiota study using next-generation sequence analysis, *Citrobacter sp.* and *Kosakonia* sp. are linked to PE of Enterobacteriaceae. The biodegradability of insects varies due to the differing chemical characteristics of different plastics. Polyethene was broken down more quickly by *Galleria mellonella* L. than polystyrene (PS). When *Tenebrio molitor* larvae were fed a combination of bran and PS, a doubling rate of PS breakdown was observed. In larvae-fed PE, beeswax can improve the gut microbiome's species richness and evenness. Like lesser waxworms, the beetle larvae can grow and degrade PS more effectively when fed PS continuously with additional nutrients. The degradation mechanism most likely plays a role in nutrient-rich diets that boost worms' gut microbiome diversity. Therefore, the optimum diet formula for combining plastics and nutrients needs to be devised, and great organisms that digest target plastics should be found with higher efficiency. According to Lwanga et al. [62], it has been demonstrated that the species *Actinobacteria (Rhodococcus justice, Mycobacterium vanbaalenii, and Streptomyces fulvissimus)* and *Firmicutes (Bacillus simplex* and *Bacillus sp.)* isolated from the gut of *Lumbricus Terrestris* degrade LDPE-MPS with high efficiency of –60% (Table 2.2).

2.3.1 Mechanism of Plastic Degradation by Insects

According to pertinent research, there are five steps in which insects break down plastics: (1) Plastics are physically chewed by the mouthparts and enter the intestinal tract; (2) plastic is adhered to and eroded by gut microbes; (3) the plastic is broken down into oligomer fragments by the oxidation or hydrolysis of enzymes provided by the gut microbiome and the host; (4) the host provides emulsifying agents that increase the efficiency of microbial and host enzymes in attacking polymers; (5) oligomer bonds are broken to form fatty acids; and the biological metabolism of insects breaks down (6) fatty acids. The role of insects' gut microbiota should be considered when looking for effective methods for plastic biodegradation. One study found that after using antibiotics to limit intestinal bacterial activity, yellow mealworms lost their capacity to break down PS, suggesting that intestinal bacteria are essential for the biodegradation of plastic [77]. After a 28-day incubation period, the isolated strain YT2 produced a biofilm on PS film. The production of C accompanied the biofilm–O polar groups, a decrease in hydrophobicity, and visible pits and cavities on the PS film surfaces. After 60 days of incubation, $7.4 \pm 0.4\%$ of the PS pieces could be broken down by a suspension culture of strain YT2. Mealworm gut bacteria play a critical role in PS biodegradation and mineralization, as evidenced by the decreased molecular weight of the remaining PS pieces and the release of water-soluble intermediates. It was also confirmed that yellow mealworms and superworms biodegrade PP through gut-microbe-dependent depolymerization [60] *Bacillus* sp. YP1

Table 2.2 The plastic-degrading insects and their ability to degrade diverse plastic materials

Insect species	Types of plastic	Source of bacteria	References
Tribolium confusum	PS, PE, and EVA (ethyl vinyl acetate)		[63]
Spodoptera frugiperda	PVC	Intestinal bacterium—strain EMBL-1	[64]
Alphitobius diaperinus	PS	Intestinal bacteria—*Pseudomonas* sp. 2 m/c	[65]
Tenebrio molitor	Polyether-PU foam	Gut microbiome—the families *Enterobacteriaceae* and *Streptococcaceae*	[66]
	PE	Gut microbiome	[67]
	PVC	Gut microbiome	[68]
	PS	Gut microbiome—*Pseudomonas* sp. EDB1, *Bacillus* sp. EDA4, and *Brevibacterium* sp. EDX	[69, 70]
	PS, LDPE	Gut microbiota and microbial functional enzymes	[71]
	LDPE	Gut microbiome—*Acinetobacter, Cloacibacterium, Corynebacterium, Curvibacter, Enhydrobacter,* and *Staphylococcus genera*	[72]
Plodia interpunctella	PE	*Meyerozyma guilliermondii* ZJC1 (MgZJC1) and *Serratia marcescens* ZJC2 (SmZJC2)	[73]
Tribolium castaneum	PS	An intestinal bacterium—*Acinetobacter* bacterium	[74]
Tenebrio obscurus	PS	Intestinal bacteria—*Enterobacteriaceae, Spiroplasmataceae,* and *Enterococcaceae*	[29]
Uloma sp.	PS	Gut microbiota	[7]
Corcyra cephalonica (Stainton)	LDPE	Gut microbiota	[75]
Plesiophthalmus davidis	PS	Gut microbiota	[76]

and *Enterobacter asburiae* YT1, two gut bacteria for PE biodegradation were identified from waxworms [77].

Furthermore, PS biodegradation and mineralization efficiency in vitro were significantly lower than in vivo, indicating that the rapid degradation of plastic in insects may be a complex mechanism that depends on the host and microbiome. According to Yang et al. (2015), the physicochemical "treatments" of chewing, ingestion, mixing with intestinal contents, and worm-secreted enzymes may be essential for quickly degrading PS in vivo. *T. molitor* produced one or more emulsifying factors (30–100 kDa) to mediate plastic bioavailability. Additionally, they showed that emulsifying factors (less than 30 kDa) released by the insect gut microbiome improved respiration on polystyrene (PS). Zhang et al. [78] found that *Aspergillus flavus,* a PE-degrading fungus in wax moth larvae, breaks down HDPE microplastics into low molecular weight pieces after 28 days [78]. The LMCOS genes Afla_053930 and Afla_006190 are up-regulated during degradation

[78]. Insect-derived bacterial and fungal enzymes probably have a direct role in the break-down of plastic by breaking down polymers into smaller molecules that cells then fully oxidize [79].

2.4   Omics to Study Microbial Degradation

The breakdown of plastics in soil and compost is more understood than the breakdown of synthetic polymers by marine microorganisms. The biodegradation of an aromatic–aliphatic copolyester blend by a marine microbial enrichment culture using metagenomics, metatranscriptomics, and metaproteomics.is well reported in the literature. Likewise it has been observed that more than six putative PETase-like enzymes and four putative MHETase-like enzymes may break down aliphatic–aromatic polymers and their breakdown products, respectively [79, 80].

2.4.1   Microbial Degradation of C–C Bond Plastics

According to Bonhomme et al. [80], the biodegradation process for PE typically entails the bio-fragmentation of the PE polymer by secreted enzymes, followed by the bio-assimilation of small lytic fragments by microbes. According to Albertsson et al. [81–83], the microbe specifically oxidizes the –C–C– groups of a long-chain backbone of PE into the –C $=$ O– (carbonyl) group, which allows the tiny aliphatic hydrocarbons to be carried straight into the cell for degradation. The primary variables influencing the pace of PS breakdown are the thickness and molecular weight of the plastics. Tischler et al. [84] showed that the actinobacterium *Rhodococcus opacus* 1CP may mineralize styrene by styrene oxide through the aerobic phenylacetic acid (PAA) pathway. This suggests that a monooxygenase-catalyzed epoxidation of the vinyl side chain degrades styrene. *Epoxystyrene isomerase* aids in converting styrene oxide to phenylacetaldehyde. Styrene monooxygenase (SMO), phenylacetaldehyde dehydrogenase (PAD), styrene oxide isomerase (SOI), and other enzymes of phenylacetate (PAA) degradation accessing the tricarboxylic acid cycle (TCA) are among the several enzymes that are involved in this metabolic pathway. *Pseudomonas* and *Xanthobacter* are two proteobacteria genera that have been shown to aerobically degrade styrene via the side-chain oxygenation pathway. Additionally, phenylacetaldehyde dehydrogenase can further oxidize phenylacetaldehyde to PAA. Because it contains Cl– , PVC's degradation route differs from PE and PP's. Chlorinated hydrocarbon oxidation is far more complicated than that of PE and PP. For instance, PVC has poor mineralization because the majority of PVC is converted to chlorinated intermediates, although PE, PP, and PS can be mineralized by *T. molitor.*

There is no report on the enzymes directly involved in the breakdown of PVC due to its hydrophobicity and chemical stability. Laccase was the only enzyme discovered

[85]. One type of oxidoreductase that has been utilized extensively to break down lignin, phenolic chemicals, and harmful pollutants is laccase (EC 1.10.3.2), which may oxidize phenolic compounds. Although the precise mechanism remained unknown, Sumathi et al. [85] demonstrated that laccase could break PVC double bonds and create new C = O bonds. Oxygen-free radicals in laccase products were thought to potentially target C–C bonds based on the breakdown paths of cellulose, lignin, and other macromolecules.

2.4.2   Microbial Degradation of Hydrolyzable Bond Plastics

To optimize the hydrophilicity of PET and the efficiency of subsequent enzymatic hydrolysis, PET and PET hydrolase can target the terminal or ring structure of the polymer chains for enzymatic hydrolysis in the ester-linked PET degradation process. According to Yoshida et al. [86], *Ideonella sakaiensis* 201-F6, a bacterium belonging to the genus Ideonella, produces PETase and MHETase, which effectively break down PET into environmentally friendly monomers, terephthalic acid, and ethylene glycol. In contrast to previous PET hydrolases, this one can break down PET with a 45–53% similarity with *actinomycete keratinase*. However, PETase's limited stability prevents it from being widely used. Interestingly, once the enzymes have been broken down, this PET hydrolase can fully dissolve PET compared to other PET hydrolases, while having 45–53% similarity with *actinomycete keratinase* [87].

2.5   Conclusion and Future Perspectives

There is a severe buildup of plastics in wild animals and ecosystems, which can endanger human health throughout the food chain. Therefore, biodegradable polymers and their ultimate breakdown without toxicity are desperately needed to address the issue of white pollution. In contrast to conventional approaches, using insects and environmental microbes to biodegrade plastics has a potential application in the industrial treatment of plastic waste. However, it is currently impractical to use insects to treat plastic trash. Environmental science encompasses both microbial and insect degradation. Depending on cost-effectiveness, enzymatic breakdown of PET may be used in the future. This review offers fresh perspectives and methods for addressing the issue of plastic pollution from a biodegradation standpoint. Further research is still advised on how insects and environmental bacteria use plastics in a biodegradable manner. Additional research on the following topics should be taken into account: (1) To enhance the library of plastic-degrading insects, more gut and environmental microorganisms with potent degrading capabilities should be screened based on the traits of these insects. (2) The amount of plastic in the food diet should be optimized further to increase the insects' effectiveness in decomposing plastic to guarantee average growth and reproduction. (3) Protein engineering and synthetic biology technology can create practical and artificial synthetic microbes by altering the enzymes that break down plastic and creating metabolic pathways. (4) Because plastics invariably impair human health and the ecological environment, a thorough toxicological analysis of plastic-degrading insects and their gut microbes should be carried out to prevent toxicological hazards. (5) Additionally, it is critical to promote the use of biodegradable plastics to eradicate the toxicity hazards at their source, particularly in areas like takeaway and e-commerce, where disposable plastic products are widely used [88, 89].

References

1. Arkatkar A, Juwarkar AA, Bhaduri S, Uppara PV, Doble M (2010) Growth of Pseudomonas and Bacillus biofilms on pretreated polypropylene surface. Int Biodeterior Biodegradation 64:530–536.
2. Artham T, Sudhakar M, Venkatesan R, Madhavan Nair C, Murty KVGK, Doble M (2009) Biofouling and stability of synthetic polymers in seawater. Int Biodeterior Biodegradation 63:884–890.
3. Kefeli, A., Razumovskii, S., Zaikov, G.Y., 1971. Interaction of polyethylene with ozone. Polym. Sci. 13 (4), 904–911.
4. Balasubramanian V, Natarajan K, Hemambika B, Ramesh N, Sumathi CS, Kottaimuthu R, Rajesh Kannan V (2010) High-density polyethylene (HDPE)-degrading potential bacteria from marine ecosystem of Gulf of Mannar India. Lett Appl Microbiol 51:205–211.
5. Balasubramanian V, Natarajan K, Rajesh Kannan V, Perumal P (2014) Enhancement of in vitro high-density polyethylene (HDPE) degradation by physical, chemical, and biological treatments. Environ Sci Pollut Res 21:12549–12562.
6. Barnes DK, Galgani F, Thompson RC, Barlaz M (2009) Accumulation and fragmentation of plastic debris in global environments. Philos Trans R Soc Lond B Biol Sci 364:1985–1998.
7. Bastioli C (2005) Handbook of biodegradable polymers. Smithers Rapra Publishing, New York.
8. Begum, M.A., Varalakshmi, B., Umamagheswari, K., 2015. Biodegradation of polythene bag using bacteria isolated from soil. Int. J. Curr. Microbiol. Appl. Sci. 4 (11), 674–680.
9. Billmeyer FW (1971) Textbook of polymer science, 2nd edn. Wiley, New York.
10. Bombelli P, Howe CJ, Bertocchini F (2017) Polyethylene bio-degradation by caterpillars of the wax moth Galleria mellonella. Curr Biol 27:292–293.
11. Briassoulis D, Aristopoulou A, Bonora M, Verlodt I (2004) Degradation characterisation of agricultural low-density polyethylene films. Biosyst Eng 88:131–143. https://doi.org/10.1016/j.biosystemseng.2004.02.010.
12. Parray, J.A., Yaseen Mir, M., Haghi, A.K. (2024). Enzymes in Valorization of Waste: Future Advancement Through the Biotechnological Revolution. In: Enzymes in Environmental Management. SpringerBriefs in Environmental Science. Springer, Cham. https://doi.org/10.1007/978-3-031-74874-5_4.
13. Parray, J.A., Yaseen Mir, M., Haghi, A.K. (2024). Enzymes Technology in Biofuel Production. In: Enzymes in Environmental Management. Springer Briefs in Environmental Science. Springer, Cham. https://doi.org/10.1007/978-3-031-74874-5_5.

14. Parray, J.A., Yaseen Mir, M., Haghi, A.K. (2024). The Potential of Enzyme Engineering to Positively Impact Environmental Sustainability. In: Enzymes in Environmental Management. SpringerBriefs in Environmental Science. Springer, Cham. https://doi.org/10.1007/978-3-031-74874-5_3.

15. Parray, J.A., Yaseen Mir, M., Haghi, A.K. (2024). Various Enzymes to Treat Resistant Pollutants in Wastewater: A Sustainable Practice for Environment. In: Enzymes in Environmental Management. SpringerBriefs in Environmental Science. Springer, Cham. https://doi.org/10.1007/978-3-031-74874-5_1.

16. Andler, Rodrigo; Tiso, Till; Blank, Lars; Andreeßen, Christina; Zampolli, Jessica; D'Afonseca, Vivian; Guajardo, Camila; Díaz-Barrera, Alvaro Current progress on the biodegradation of synthetic plastics: from fundamentals to biotechnological applications . Reviews in Environmental Science and Bio/Technology (2022).

17. Parray, J.A., Yaseen Mir, M., Shafi N., Haghi, A.K. (eds) (2025). Ozone Technology for Food Processing and Preservation . In: Synthesis Lectures on Chemical Engineering and Biochemical Engineering . Springer, Cham. ISBN: 978-3-031-81460-0.

18. Tribedi P, Sil AK (2013) Low-density polyethylene degradation by *Pseudomonas* sp. AKS2 biofilm. Environ Sci Pollut Res Int 20:4146–4153.

19. Awasthi, S., Srivastava, P., Singh, P., Tiwary, D., Mishra, P.K., 2017. Biodegradation of thermally treated high-density polyethylene (HDPE) by Klebsiella pneumoniae CH001. 3 Biotech 7 (5), 332.

20. Parray, J.A., Yaseen Mir, M., Shafi N., Haghi, A.K. (eds) (2025). Microplastics Pollution Control in Water Systems. Springer, Cham. ISBN: 978-3-031-74398-6.

21. Byuntae L, Anthony LP, Alfred F, Theodore BB (1991) Biodegradation of degradable plastic polyethylene by Phanerocheate and Streptomyces species. Appl Environ Microbiol 3:678–688.

22. Chatterjee S, Roy B, Roy D, Banerjee R (2010) Enzyme-mediated biodegradation of heat treated commercial polyethylene by Staphylococcal species. Polym Degrad Stab 95:195–200.

23. Chen, Q., Lv, W., Jiao, Y., Liu, Z., Li, Y., Cai, M., Wu, D., Zhou, W., Zhao, Y., 2020a. Effects of exposure to waterborne polystyrene microspheres on lipid metabolism in the hepatopancreas of juvenile redclaw crayfish, Cherax quadricarinatus. Aquat. Toxicol. 224, 105297.

24. Chen, Y., Wen, D., Pei, J., Fei, Y., Ouyang, D., Zhang, H., Luo, Y., 2020b. Identification and quantification of microplastics using Fourier-transform infrared spectroscopy: Current status and future prospects. Curr. Opin. Environ. Sci. Health. 18, 14–19.

25. Chiellini E, Corti A, Swift G (2003) Biodegradation of thermally-oxidized fragmented low-density polyethylenes. Polym Degrad Stab 81:341–351.

26. Corami, F., Rosso, B., Bravo, B., Gambaro, A., Barbante, C., 2020. A novel method for purification, quantitative analysis and characterization of microplastic fibers using Micro-FTIR. Chemosphere 238, 124564.

27. Cornell JH, Kaplan AM, Rogers MR (1984) Biodegradability of photooxidized polyalkylenes. J Appl Polym Sci 29:2581–2597.

28. Curlee TR, Das S (1991) Identifying and assessing targets of opportunity for plastics recycling. Resour Conserv Recycl 5:343–363.

29. Peng, B. Y., Su, Y., Chen, Z., Chen, J., Zhou, X., Benbow, M. E., et al. (2019). Biodegradation of Polystyrene by Dark (Tenebrio obscurus) and Yellow (Tenebrio molitor) Mealworms (Coleoptera: Tenebrionidae). *Environ. Sci. Technol.* 53, 5256–5265. https://doi.org/10.1021/acs.est.8b06963.

30. Danso D, Chow J, Streit WR (2019) Plastics: environmental and biotechnological perspectives on microbial degradation. Appl Environ Microbiol 85:1–14.

31. Danso, D., Chow, J., Streit, W.R., 2018. Plastics: Environmental and biotechnological perspectives on microbial degradation. Appl. Environ. Microbiol. 85 (19), e01095–19.

32. Das MP, Kumar S (2014) Microbial deterioration of low density polyethylene by Aspergillus and Fusarium sp. Int J ChemTech Res 6:299–305.
33. Das, K., Mukherjee, A.K., 2005. Characterization of biochemical properties and biological activities of biosurfactants produced by Pseudomonas aeruginosa mucoid and non-mucoid strains isolated from hydrocarbon-contaminated soil samples. Appl. Microbiol. Biotechnol. 69 (2), 192–199.
34. Das, M.P., Kumar, S., 2015. An approach to low-density polyethylene biodegradation by Bacillus amyloliquefaciens. 3 Biotech 5 (1), 81–86.
35. de Souza Machado AA, Kloas W, Zarfl C, Hempel S, Rillig MC (2018) Microplastics as an emerging threat to terrestrial ecosystems. Glob Change Biol 24:1405–1416.
36. Deguchi T, Kitaoka Y, Kakezawa M, Nishida T (1998) Purification and characterization of a nylon-degrading enzyme. Appl Environ Microbiol 64:1366–1371.
37. Ehara K, Iiyoshi Y, Tsutsumi Y, Nishida T (2000) Polyethylene degradation by manganese peroxidase in the absence of hydrogen peroxide. J Wood Sci 46:180–183.
38. Koutny M, Lemaire J, Delort AM (2006) Biodegradation of polyethylene films with pro-oxidant additives. Chemosphere 64:1243–1252.
39. Koutny M, Sancelme M, Dabin C, Pichon N, Delort A, Lemaire J (2006) Acquired biodegradability of polyethylenes containing pro-oxidant additives. Polym Degrad Stab 91:1495–1503.
40. Krueger MC, Harms H, Schlosser D (2015) Prospects for microbiological solutions to environmental pollution with plastics. Appl Microbiol Biotechnol 99:8857–8874.
41. Krupp LR, Jewell WJ (1992) Biodegradability of modified plastic films in controlled biological environments. Environ Sci Technol 26:193–198.
42. Krzan, A., Hemjinda, S., Miertus, S., Corti, A., Chiellini, E., 2006. Standardization and certification in the area of enivironmentaly degradable plastics. Polym. Degrad. Stabil. 91 (12), 2819–2833.
43. Kumar, S., Maiti, P., 2016. Controlled biodegradation of polymers using nanoparticles and its application. RSC Adv. 6 (72), 67449–67480.
44. Lebreton L, Slat B, Ferrari F, Sainte-Rose B, Aitken J, Marthouse R, Hajbane S, Cunsolo S, Schwarz A, Levivier A, Noble K, Debeljak P, Maral H, Schoeneich-Argent R, Brambini R, Reisser J (2018) Evidence that the Great Pacific Garbage Patch is rapidly accumulating plastic. Sci Rep 8:4666.
45. Li, J., Guo, S., Li, X., 2005. Degradation kinetics of polystyrene and EPDM melts under ultrasonic irradiation. Polym. Degrad. Stabil. 89 (1), 6–14.
46. Lin, Y.-H., Yen, H.-Y., 2005. Fluidised bed pyrolysis of polypropylene over cracking catalysts for producing hydrocarbons. Polym. Degrad. Stabil. 89 (1), 101–108.
47. Usha R, Sangeetha T, Palaniswamy M (2011) Screening of polyethylene degrading microorganisms from garbage soil. Libyan Agric Res Cent J Int 2:200–204.
48. Yamada-Onodera K, Mukumoto H, Katsuyaya Y, Saiganji A, Tani Y (2001) Degradation of polyethylene by a fungus, *Penicillium simplicissimum* YK. Polym Degrad Stab 72:323–327.
49. Esmaeili A, Pourbabaee AA, Alikhani HA, Shabani F, Esmaeili E (2013) Biodegradation of low-density polyethylene (LDPE) by mixed culture of *Lysinibacillus xylanilyticus* and *Aspergillus niger* in soil. PLoS ONE 8:717–720.
50. Lucas, N., Bienaime, C., Belloy, C., Quenedecu, M., Silvestre, J.-E., Saucedo, N., 2008. Polymer biodegradation: mechanisms and estimation techniques. Chemosphere 73, 429–442.
51. Zeenat, A. Elahi, Dilara Elahi, A., Bukhari, D. A., Shamim, S., & Rehman, A. (2021). Plastics degradation by microbes: A sustainable approach. *Journal of King Saud University - Science*, 33(6), 101538. https://doi.org/10.1016/j.jksus.2021.101538.

52. Palm, G.J., Reisky, L., Böttcher, D. *et al.* Structure of the plastic-degrading *Ideonella sakaiensis* MHETase bound to a substrate. *Nat Commun* **10**, 1717 (2019). https://doi.org/10.1038/s41 467-019-09326-3.

53. Thomas, B., Olanrewaju-Kehinde, D., Popoola, O., James, E., 2015. Degradation of plastic and polythene materials by some selected microorganisms isolated from soil. World Appl. Sci. J. 33 (12), 1888–1891.

54. Skariyachan, S., Setlur, A.S., Naik, S.Y., Naik, A.A., Usharani, M., Vasist, K.S., 2017. Enhanced biodegradation of low and high-density polyethylene by novel bacterial consortia formulated from plastic-contaminated cow dung under thermophilic conditions. Environ. Sci. Pollut. Res. 24, 8443–8457.

55. Muhonja, C.N., Makonde, H., Magoma, G., Imbuga, M., 2018. Biodegradability of polyethylene by bacteria and fungi from Dandora dumpsite Nairobi-Kenya. PLoS ONE 13, (7) e0198446.

56. Munir, E., Harefa, R.S.M., Priyani, N., Suryanto, D., 2018. Plastic degrading fungi Trichoderma viride and Aspergillus nomius isolated from local landfill soil in Medan. IOP Conf. Series: Earth and Environmental. Science 126, 012145.

57. . Park, S.Y., Kim, C.G., 2019. Biodegradation of micro-polyethylene particles by bacterial colonization of a mixed microbial consortium isolated from a landfill site. Chemosphere 222, 527–533.

58. Taghavi, N., Singhal, N., Zhuang, W.-Q., Baroutian, S., 2021. Degradation of plastic waste using stimulated and naturally occurring microbial strains. Chemosphere 263, 127975.

59. Maroof, L., Khan, I., Yoo, H.S., Kim, S., Park, H.-T., Ahmad, B., Azam, S., 2021. Identification and characterization of low density polyethylene-degrading bacteria isolated from soils of waste disposal sites. Environ. Eng. Res. 26, (3) 200167.

60. Yang, Y., Liu, W., Zhang, Z., Grossart, H.-P., Gadd, G.M., 2020. Microplastics provide new microbial niches in aquatic environments. Appl. Microbiol. Biotechnol. 104, 6501–6511.

61. Yang, Z., Lü, F., Zhang, H., 61 Wang, W., Shao, L., Ye, J., He, P., 2021. Is incineration the terminator of plastics and microplastics? J. Hazard. Mater. 491, 123429.

62. Lwanga EH, Thapa B, Yang X, Gertsen H, Salánki T, Geissen V, Garbeva P (2018) Decay of low-density polyethylene by bacteria extracted from earthworm's guts: a potential for soil restoration. Sci Total Environ 624:753–757.

63. Abdulhay, H. (2020). Biodegradation of plastic wastes by confused flour beetle Tribolium confusum Jacquelin du Val larvae. *Asian J. Agric. Biol.* 8, 201–206. https://doi.org/10.35495/ajab. 2019.11.515.

64. Zhu, P., Shen, Y., Li, X., Liu, X., Qian, G., and Zhou, J. (2022). Feeding preference of insect larvae to waste electrical and electronic equipment plastics. *Sci. Total Environ.* 807, 151037. https://doi.org/10.1016/j.scitotenv.2021.151037.

65. Cucini, C., Funari, R., Mercati, D., Nardi, F., Carapelli, A., and Marri, L. (2022). Polystyrene shaping effect on the enriched bacterial community from the plastic-eating Alphitobius diaperinus (Insecta: Coleoptera). *Symbiosis* 86, 305–313. https://doi.org/10.1007/s13199-022-008 47-y.

66. Liu, J., Liu, J., Xu, B., Xu, A., Cao, S., Wei, R., et al. (2022). Biodegradation of polyetherpolyurethane foam in yellow mealworms (Tenebrio molitor) and effects on the gut microbiome. *Chemosphere* 304, 135263. https://doi.org/10.1016/j.chemosphere.2022.135263.

67. Bulak, P., Proc, K., Pytlak, A., Puszka, A., Gawdzik, B., and Bieganowski, A. (2021). Biodegradation of different types of plastics by tenebrio molitor insect. *Polymers* 13, 3508. https://doi. org/10.3390/polym13203508.

68. Peng, B.-Y., Chen, Z., Chen, J., Yu, H., Zhou, X., Criddle, C. S., et al. (2020). Biodegradation of Polyvinyl Chloride (PVC) in Tenebrio molitor (Coleoptera: Tenebrionidae) larvae. *Environ. Int.* 145, 106106. https://doi.org/10.1016/j.envint.2020.106106.

69. Arunrattiyakorn, P., Ponprateep, S., Kaennonsang, N., Charapok, Y., Punphuet, Y., Krajangsang, S., et al. (2022). Biodegradation of polystyrene by three bacterial strains isolated from the gut of Superworms (Zophobas atratus larvae). *J. Appl. Microbiol.* 132, 2823–2831. https://doi.org/10.1111/jam.15474.

70. Yang X-G, Wen P-P, Yang Y-F, Jia P-P, Li W-G and Pei D-S (2023) Plastic biodegradation by *in vitro* environmental microorganisms and *in vivo* gut microorganisms of insects. *Front. Microbiol.* 13:1001750. https://doi.org/10.3389/fmicb.2022.1001750.

71. Peng, B.-Y., Sun, Y., Wu, Z., Chen, J., Shen, Z., Zhou, X., et al. (2022). Biodegradation of polystyrene and low-density polyethylene by Zophobas atratus larvae: Fragmentation into microplastics, gut microbiota shift, and microbial functional enzymes. *J. Cleaner Prod.* 367, 132987. https://doi.org/10.1016/j.jclepro.2022.132987.

72. Latour, S., Noël, G., Serteyn, L., Sare, A. R., Massart, S., Delvigne, F., et al. (2021). Multi-omics approach reveals new insights into the gut microbiome of *Galleria mellonella* (Lepidoptera: Pyralidae) exposed to polyethylene diet. bioRxiv.2021.2006.2004.446152. https://doi.org/10.1101/2021.06.04.446152.

73. Lou, H., Fu, R., Long, T., Fan, B., Guo, C., Li, L., et al. (2022). Biodegradation of polyethylene by Meyerozyma guilliermondii and Serratia marcescens isolated from the gut of waxworms (larvae of Plodia interpunctella). *Sci. Total Environ.* 853, 158604. https://doi.org/10.1016/j.scitotenv.2022.158604.

74. Wang, L., Peng, Y., Xu, Y., Zhang, J., Liu, C., Tang, X., et al. (2022). Earthworms' degradable bioplastic diet of polylactic acid: easy to break down and slow to excrete. *Environ. Sci. Technol.* 56, 5020–5028. https://doi.org/10.1021/acs.est.1c08066.

75. Kesti, S., and Sharana, S. (2019). First report on biodegradation of low density polyethylene by rice moth larvae, Corcyra cephalonica (Stainton). *Holistic Appr. Environ.* 9, 79–83. https://doi.org/10.33765/thate.9.4.2.

76. Woo, S., Song, I., and Cha, H. J. (2020). Fast and Facile Biodegradation of Polystyrene by the Gut Microbial Flora of *Plesiophthalmus davidis* Larvae. *Appl. Environ. Microbiol.* 86, e01361-20. https://doi.org/10.1128/AEM.01361-20.

77. Yang, J., Yang, Y., Wu, W. M., Zhao, J., and Jiang, L. (2014). Evidence of polyethylene biodegradation by bacterial strains from the guts of plastic-eating waxworms. *Environ. Sci. Technol.* 48, 13776–13784. https://doi.org/10.1021/es504038a.

78. Zhang, J., Gao, D., Li, Q., Zhao, Y., Li, L., Lin, H., et al. (2020). Biodegradation of polyethylene microplastic particles by the fungus Aspergillus flavus from the guts of wax moth Galleria mellonella. *Sci. Total Environ.* 704, 135931. https://doi.org/10.1016/j.scitotenv.2019.135931.

79. Amobonye, A., Bhagwat, P., Singh, S., and Pillai, S. (2021). Plastic biodegradation: Frontline microbes and their enzymes. *Sci. Total Environ.* 759, 143536. https://doi.org/10.1016/j.scitotenv.2020.143536.

80. Bonhomme S, Cuer A, Delort A, Lemaire J, Sancelme M, Scott G (2003) Environmental biodegradation of polyethylene. Polym Degrad Stab 81:441–452.

81. Albertsson AC (1980) The shape of the biodegradation curve for low and high density polyethenes in prolonged series of experiments. Eur Polym J 16:623–630.

82. Albertsson AC, Barenstedt C, Karlsson S, Lindberg T (1995) Degradation product pattern and morphology changes as means to differentiate abiotically and biotically aged degradable polyethylene. Polymer 36:3075–3083.

83. Albertsson AC, Karlsson S (1990) The influence of biotic and abiotic environments on the degradation of polyethylene. Prog Polym Sci 15:177–192.

84. Tischler, D., Eulberg, D., Lakner, S., Kaschabek, S. R., Van Berkel, W. J. H., and Schlömann, M. (2009). Identification of a novel self-sufficient styrene monooxygenase from Rhodococcus opacus 1CP. *J. Bacteriol.* 191, 4996–5009. https://doi.org/10.1128/JB.00307-09.

85. Sumathi, T.; Viswanath, B.; Sri Lakshmi, A.; SaiGopal, D.V.R. Production of Laccase by Cochliobolus sp. Isolated from Plastic Dumped Soils and Their Ability to Degrade Low Molecular Weight PVC. Biochem. Res. Int. 2016, 2016, 9519527.

86. Yoshida, S., Hiraga, K., Takehana, T., Taniguchi, I., Yamaji, H., Maeda, Y., et al. (2016). A bacterium that degrades and assimilates poly(ethylene terephthalate). *Science* 351, 1196–1199. https://doi.org/10.1126/science.aad6359.

87. Wei, R., Song, C., Gräsing, D., Schneider, T., Bielytskyi, P., Böttcher, D., et al. (2019). Conformational fitting of a flexible oligomeric substrate does not explain the enzymatic PET degradation. *Nature Communications* 10, 5581. https://doi.org/10.1038/s41467-019-13492-9.

88. Lomonaco, T., Manco, E., Corti, A., La Nasa, J., Ghimenti, S., Biagini, D., Di Francesco, F., Modugno, F., Ceccarini, A., Fuoco, R., Castelvetro, V., 2020. Release of harmful volatile organic compounds (VOCs) from photo-degraded plastic debris: A neglected source of environmental pollution. J Hazard. Mater. 394, 122596.

89. Kundungal, H., Gangarapu, M., Sarangapani, S., Patchaiyappan, A., and Devipriya, S. P. (2021a). Role of pretreatment and evidence for the enhanced biodegradation and mineralization of low-density polyethylene films by greater waxworm. *Environ. Technol.* 42, 717–730. https://doi.org/10.1080/09593330.2019.1643925.

90 Yang Y, Yang J, Wu WM, Zhao J, Song Y, Gao L, Yang R, Jiang L (2015c) Biodegradation and mineralization of polystyrene by plastic-eating mealworms: Part 1. Chemical and physical characterization and isotopic tests. Environ Sci Technol 49:12080–12086. https://doi.org/10.1021/acs.est.5b02661

Mechanism of Actions of Microbes for Degradation of Plastics

3

3.1 Microbial Interaction and Plastic Biodegradation

The disposal of plastic trash may benefit from the presence of microorganisms that may break down polymers. Japanese researchers examined ponds filled with wastewater from a nylon mill in 1975; they found a type of Flavobacterium that breaks down the linear dimer of 6-aminohexanoate and other unavoidable leftovers from the production of nylon-6. In sludge, the ND-10 and ND-11 strands of *Pseudomonas* sp. can break down nylon 4 (polybutyrolactam), producing GABA (γ-aminobutyric acid) as a byproduct [1, 2]. Two species of the Ecuadorian fungus *Pestalotiopsis,* found at the bottom of landfills, are among the many soil fungal species that can both aerobically and anaerobically degrade polyurethane. Styrene is utilized as a carbon source and is broken down by methanogenic microbial consortia. Microbial communities extracted from starch-mixed soil samples like *Aspergillus fumigatus* effectively degrades plasticized PVC. *Phanerochaete chrysosporium* was cultivated on PVC in mineral salt agar. PVC is also efficiently broken down by *P. chrysosporium, Lentinus tigrinus, A. niger,* and *A. sydowii.* It has been discovered that *Acinetobacter* partially breaks down low-molecular-weight polyethene oligomers. In less than three months, *Pseudomonas fluorescent and Sphingomonas* may break down more than 40% of the weight of plastic bags when combined [3]. Onboard space stations have discovered dangerous moulds that break down rubber into a form that may be consumed. On artificial polymer objects found in museums and archaeological sites, various yeast, bacterial, algal, and lichen species have flourished. Sargasso Sea bacteria have been shown to devour a variety of plastics in the plastic-polluted seas. However, it is unclear how well these bacteria work. Microbes that consume plastic have also been discovered in landfills. PET can be broken down by Nocardia using an esterase enzyme [4]. It has been found that the polycarbonate plastic present in CDs is consumed by the Belizean fungus *Geotrichum candidum.*

© The Author(s), under exclusive license to Springer Nature Switzerland AG 2025
J. A. Parray and W.-J. Li, *Microbial and Enzyme-Based Technology for Plastic Biodegradation*, Synthesis Lectures on Chemical Engineering and Biochemical Engineering, https://doi.org/10.1007/978-3-031-84437-9_3

3.2   Mechanisms of Plastic Biodegradation

After microorganisms depolymerize plastics or microplastics, the bacteria release free radicals and extracellular enzymes that catalytically break down the biodegraded polymers into smaller components. Polymers are typically broken down into shorter chains or smaller molecules, such as oligomers, dimers, and monomers, during the microbiological degradation of plastics. Depolymerization of sufficiently small polymers results in monomers that may be required for microbial growth and uptake via semipermeable membranes and, ultimately, for mineralization in cells. To create biomass for energy, the cell monomers are then mineralized into CO_2, H_2O (in aerobic settings) or CO_2, H_2O, and CH_4 (in anaerobic conditions). The first stage of microbial degradation of plastics, which serves as the foundation for enzymatic breakdown, is microbial colonization on plastic surfaces by adhesion or exposure. The enzyme and polymer matrix combination is used in the second phase, hydrolysis, followed by catalytic hydrolysis and cracking. The hydrolysis of organic materials is catalyzed by the attached enzyme, a hydrolase [5–7]. Microorganisms' external and intracellular enzymes are involved in two vital functions. The polymer chain undergoes hydrolytic cleavage due to the action of extracellular enzymes, including hydrolases and depolymerase.

Enzyme amplification attacks the polymer chain, producing tiny oligomers or monomers. Furthermore, following their metabolism, tiny oligomers or monomers might be absorbed by enzymes within the cell [8]. Weight loss and adding functional groups are indicators of enzymatic degradation instead of oxidative breakdown for big plastics or plastic pieces. The weight loss of the polymers after biodegradation is one of the most significant changes in MPs. This is the initial weight of plastics or microplastics and the weight attained following microbe contact [9–11]. The environment benefits from something made by microbial metabolism and breakdown. One crucial component influencing biodegradability is polymer shape, such as crystallinity, which has a significant impact on the biodegradation rate.

This is because the enzyme primarily targets the polymer's amorphous domain, which has loosely structured molecules that are easier to break down [12]. Thus, microbial enzymes contribute to bio-transformation and biodegradation processes, play a critical role in environmental remediation and ecological health through biodegradation, and combine the benefits of efficiency and functionality to make bio-catalytic processes more competitive, safe, clean, and environmentally friendly [13].

3.2.1   Microorganisms and Their Enzymes in Plastic Degradation

Three technological outcomes can primarily determine the evidence of degradation: (1) structural changes in plastics or microplastics, (2) physical mass loss of plastics or MPs, and (3) the production of plastics or MP metabolites. These combined findings are the most substantial evidence for plastics/MP biodegradation [3, 14, 15]. In addition to PE (polyethene), PP (polypropylene), PS (polystyrene), PVC (polyvinyl chloride), PET (polyethene terephthalate), PC (polycarbonate), PMMA (polymethyl methacrylate), and PU (polyurethane), the principal molecular chain of nonhydrolyzable plastics is composed solely of C–C bonds [16–18]. Before being used by microorganisms, nonhydrolyzable polymers are frequently transformed by bacteria or fungi into tiny molecules of organic matter through redox processes [19]. The hydrolyzable polymers PA and PET break down rather quickly.

An increasing number of examples demonstrate that microbes can also dissolve plastic components [20]. Few bacteria and fungi have been shown to partially break down PET at, and the majority of bacterial isolates that are capable of doing so belong to the Gram-positive phylum Actinobacteria [21], which includes the genera *Thermobifida* and *Thermomonospora* [22]. PET metabolism has been linked to several signature enzymes in the database, including PETase, MHETase, TPA dioxygenase, and PCA dioxygenase. A novel bacterium called *Ideonella sakaiensis* 201-F has been shown in one case study report to use PET as its primary source of carbon and energy and to alter the shape of PET film. This bacterium was found to be unique because it can only degrade polymers. PET can be hydrolyzed by the strain's production of two enzymes i.e. PET hydrolase and PETase, which effectively breaks down PET into environmentally safe monomers [23–25]. Similar to certain bacteria, strains of *I. sakaiensis* release enzymes that hydrolyze microplastics and cause them to mineralize within cells [26]. The procedure consists of two stages: the strain produces two particular enzymes, MHETase and PETase. Nevertheless, it is unknown exactly how PETase binds. Furthermore, *Pseudomonas monteilii S17 and Pseudomonas nitroreducens* S8 can use PET as a carbon source by colonizing the PET surface as a transparent biofilm layer with a constant population density. Strong synergistic effects were seen when *P. nitroreducens* S8 and *P. monteilii* S17, which create a PET surface that is very accessible for PET hydrolases, were used together. A synthetic hydrophobic polymer with a high molecular weight, PS (polystyrene), is complex for microbes to attack and destroy. PS biodegradation is extremely little. Nonetheless, the study found that mealworms' intestinal tracts can degrade PS. Two operational taxonomic units (OTUs) in the mealworm gut microbiome, *Citrobacter* sp. and *Kosakonia* sp. were discovered to have strong relationships with PE and PS, suggesting that they could degrade chemically different plastics [27–29].

3.2.2   Effects of Microbial Activity on PE

By tracking seven distinct attributes related to the degree of biodegradation of the polymer—functional groups on the surface, hydrophobicity/hydrophilicity, crystallinity, molecular weight distribution, surface topography, mechanical properties, and mass balance—the impact of microbial colonization on the PE surface was investigated. The techniques utilized to examine these alterations have been well addressed in other research. The development of different functional groups on the surface of PE during abiotic and biotic oxidation by thermo-UV treatment and microbial degradation is investigated using FTIR spectroscopy [30–35].

Infrared absorbance at 1710–1715 cm^{-1} (corresponding to carbonyl group), 1640 cm^{-1}, and 830–880 cm^{-1} (corresponding to –C=C–) increased when PE was treated with UV and nitric acid, for instance. This absorbance decreased following incubation with microbial consortia. Likewise, Harshvardhan and Jha [36] used FT-IR spectra to demonstrate PE biodegradation with a rise in the index of the vinyl bond, the carbonyl bond, and the keto carbonyl bond. Because oxidized groups increase hydrophilicity, which leads to the efficient attachment of microorganisms to the PE surface and promotes biodegradation, these functional groups at the PE surface are considered significant [2, 21]. The contact angle between the surface and water is typically used to determine hydrophilicity. The oxidized PE surface is extremely hydrophilic when the contact angle with water is slight [37–39]. FTIR analysis and differential scanning calorimetry (DSC) are used to quantify crystallinity, another crucial metric used to forecast the degree of biodeterioration of the polymer. In general, microbes may readily access and break down the amorphous zone, which causes an initial rise in crystallinity [5, 40–44]. Microorganisms will begin breaking down the crystalline zone and increasing the percentage of giant crystals after attacking (or residing in) the amorphous regions [2, 45].

Time-of-flight mass spectrometry (TOF-MS) analysis and size exclusion chromatography provide insight into the molecular weight distribution of the PE during biodegradation. Following the first breakdown of low-molecular-weight chains, an increase in the average molecular weight is observed [5, 41]. When examining the surface topography of PE films during biodegradation, SEM and AFM investigations are frequently used. Following microbial incubation, biofilm formation is typically seen on the polymer surface [14, 35, 38, 46–50] and hyphal structure penetration [44]. Under SEM, structural alterations in the development of pits, holes, and erosions have been seen signifying the surface degradation of PE.

Oxidation-induced changes in crystallinity and average molecular weight alter PE's chemical and mechanical properties. A preferred method for examining alterations in a polymer's mechanical characteristics is the universal mechanical testing system (UMTS). However, this approach tends to understate the damage that the microbe produce to the local surface. Since microorganisms use PE as a carbon source and convert it to CO_2 during respiration, the amount of polymer consumed can be connected to the measured amount of CO_2 emitted. According to specific research, samples' weight decreased as determined by CO_2 emissions from the samples or gravimetric measurements [50–55].

The overall degradation of the polymer and its rate are determined by measuring the samples' progressive CO_2 emissions. In addition to nonhydrolyzable PS, PE biodegradation is extremely difficult. PE is a thermoplastic hydrocarbon polymer commonly made and sold globally as a synthetic polymer. It has electrical insulation and corrosion resistance [32]. PE has extremely resilient characteristics because of its nonhydrolyzable covalent bonds and carbon–carbon backbone. Regarding function and natural synergy, microbial consortiums may be a reasonably successful option for PE decomposition [9].

Bacillus, Mycobacterium, Nocardia, and Pseudomonas species are some examples of bacteria that break down PE [42]. A bacterial consortium has been found to have the ability to create a thick biofilm on the worn PE surface, altering the surface's topography and rheology. According to the mesophilic bacterial consortium isolated from landfill debris, the mean MP particle diameter and dry weight after 60 days were reduced by 22.8 and 14.7%, respectively, by the *Bacillus* sp. and *Paenibacillus* sp. combinations [5].

The two most widely used PE polymers are high-density polyethene (HDPE) and low-density polyethene (LDPE), with LDPE emerging as one of the contemporary polymers' most significant plastic grades. HDPE is comparatively rugged to break down in terms of microbial breakdown because of its production function [42, 43]. In the biodegradation process, fungi outperformed bacteria in the decomposition of PE, and fungi-treated LDPE films showed a more significant weight loss than bacteria, according to a comparison of fungi and bacteria isolated from identical waste [44]. LDPE can be effectively broken down by bacterial and fungal strains acting alone or in combination [45]. Microorganisms isolated from marine environments to grow bacteria use the carbon in LDPE.

LDPE films with four bacterial isolates showed dry weight losses of 1, 0, 78, 22, and 0.46% after 30 days of incubation and 1, 1, 72, and 0.97% after 90 days. Weight loss and carbon mineralization calculations of LDPE film are used to assess how well LDPE works as a carbon source for bacterial growth. The bacterially treated LDPE films' FTIR spectra showed new peak formation and shift for the functional groups following a 90-day incubation period. All four of the marine bacteria-treated LDPE films had surface degradation, fragility, broken layers, cracks, and scratches, according to SEM pictures [42]. *Penicillium oxalicum* NS4 (KU559906) and *Penicillium chrysogenum* NS10 (KU559907) are known fungal strains that break down plastics. They have been isolated from fungi and have been shown to degrade HDPE and LDPE while also causing observable morphological changes on the plastic sheet [46]. The most common fungal group is the strain *Aspergillus*, which has demonstrated the ability to break down solid plastic. The *Aspergillus* genus, which includes *A. clavatus, A. fumigatus, and A. niger,* breaks down synthetic polymers like PE and PP [33]. HDPE can be broken down by *Aspergillus niger* (ITCC No. 6052), which was isolated from a plastic waste wasteland [47]. Gaining a thorough understanding of *A. niger* and its enzyme system will aid in comprehending its function in HDPE biodegradation. LDPE and HDPE are broken down by *Aspergillus* species, which include *Rhizopus oryzae* strains NS5 and *A. niger*, respectively. Furthermore, two strains of Penicillium (*Penicillium oxalicum* NS4 and *Penicillium chrysogenum* NS10) demonstrated characteristics for both LDPE and HDPE [48]. Numerous factors affected the weight loss; the most notable weight loss was in LDPE and HDPE foils, films, or strips.

3.2.3 Microbial Degradation of Toxic Components in Commonly Used Plastics

Various polymers infused with different additives and fillers have been produced in vast quantities since the 1950s, when plastics were first manufactured in large quantities. Two common plastics that potentially have adverse effects are PAEs (athletes) and BPA (bisphenol A). BPA poisons the majority of marine creatures and vertebrates. Because BPA shares structural similarities with diethylstilbestrol (DES), it may be carcinogenic. As a result, exposure to plastic particles or even more miniature stages increases toxicity [49]. The biodegradation of plastics involves the hazardous chemicals PAEs and BPA, which are among the primary constituents of plastics. Bacteria, including Pseudomonas, *Arthrobacter, Rhodococcus, Bacillus, Mycobacterium, Delfia, and Gordonia,* are used in most investigations on the biodegradation of PAE [13].

PAEs are broken down by microbes by the hydrolysis of ester bonds, producing monoesters, to further end products i.e. alcohol and phthalic acid (PA) than it activates the Krebs cycle by producing short-chain acids [50]. The enzymes that aid in the biodegradation of PAEs are oxygenase and hydrolase. Esterase, an enzyme with a pH range of 7–10 and a temperature range of 30 to 70 °C, exhibits the highest activity in breaking down phthalates [51]. It was revealed that decrease in weight is predominant in case pretreatments like temperature and addition of hydrolases in an estuary waste electronics dismantling area. *Bacillus sp.* GZB-isolated as facultative anaerobic BPA-degrading bacteria shows promising activity in BPA biodegradation in water sediments.

3.2.4 Microbial Degradation of Commonly Used Biodegradable Plastics

Biodegradable plastics, a new type of plastic, are degrade more quickly than high polymers. When compared to high polymers, biodegradable plastics (BPs) like polycaprolactone (PCL), polybutylene succinate-co-adipate (PBSA), polybutylene succinate (PBS), and polylactic acid (PLA) are more likely to degrade and get past the problems and environmental pollution that come with synthetic plastics. Significant degradation is possible for a few biodegradable polymers used in composting, including PCL, PLA, PBS, and polyhydroxybutyrate (PHB). Importantly, their chemical structure contains ester bonds that allow microbial enzyme systems to break them down [15]. Thermophilic actinomycetes that are active against PHB, PCL, or PES were isolated from various Taiwanese habitats, including compost [54]. Two thermophilic bacteria isolated from compost showed effective synergy when exposed to PCL wastes at 50 °C, speeding up PCL breakdown and significantly increasing the amount of disintegrated polymer [55]. Lipase can hydrolyze PCL into little molecules because it is a thermoplastic crystalline

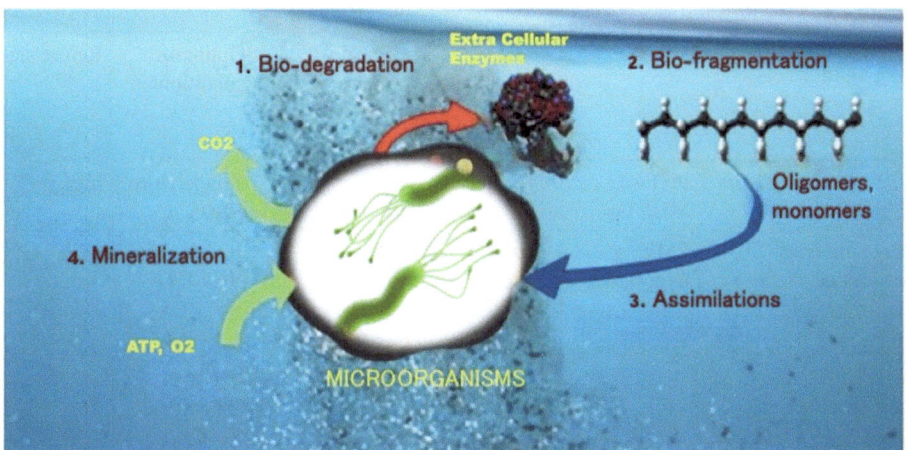

Fig. 3.1 Mechanism of abiotic and biotic degradation of polyethylene

polyester, and microbes can further absorb it. For example, *Aspergillus* sp. strain ST-01, isolated from soil and incubated for six days at 50 °C by thermotolerant PCL-degrading bacteria, completely broke down PCL. Aspergillus species secrete the enzymes catalase and protease strain ST-01. Marine organisms identified from deep-sea sediments, including *Pseudomonas, Alcanivorax, and Tenacibaculum*, can effectively degrade PCL [56–60] (Fig. 3.1).

3.3 Microbial Enzymes Involved in Plastic Biodegradation

Plastic biodegradation involves several biotic and abiotic components [29, 40, 56, 61, 62]. The bulk polymer fragments result from abiotic and microbiological support, increasing the number of surfaces available for biodegradation. Additionally, the PE thermoplastic can be broken down by several lignin-degrading enzymes [63] and several extracellular enzymes [63–65]. The polymer can be taken into the cells for further metabolism following its initial cleavage into oligomers of 10–50 carbon atoms. The lack of hydrolyzable functional groups in the backbone of PE limits its biodegradation. Using substrate peroxidation, the carbonyl and hydroxyl groups produced by different pretreatments, like thermo-UV irradiation or the addition of oxidizing agents, could be used to promote biodegradation [35, 54] (Table 3.1).

The biodegradation of PE has been facilitated by microbial enzymes that can break down lignin polymers with oxidizable C–C linkages [36]. These enzymes include laccases (EC 1.10.3.2.), manganese peroxidase (MnP, EC 1.11.1.13), and lignin peroxidases

Table 3.1 Enzymes and degradation factors of degrading bacteria

Source	Enzyme	Major mechanism of degradation	Plastics/ microplastics	Optimum conditions	References
Ideonella sakaiensis 201-F	PETase; MHETase	Hydrolysis	PET	Temperature 70–75 °C	[66]
Pseudomonas, arthrobacter	PME hydrolases	Hydrolysis and oxidation	PVC, PP, PE, PS (PAEs)	Temperature 30–70 °C	[67]
Bacillus sp. GZB	Spore-laccase	Expression of different functional genes	PC (BPA)	Adding electron donors and co-substrates	[68]
Aspergillus sp. strain ST-01	Catalase, protease	Colonization	PCL	Temperature 50 °C	[69]
Thermomonospora curvata (cutinase homolog from leaf-branch compost)	LC-cutinase	Hydrolysis	PET	Temperature 50 °C	[70]
Thermophilic, alkaliphilic, halophilic, and psychrophilic bacteria	Bacteriophilic enzyme	Hydrolysis	Various plastics	Salt, low, or high pH temperatures	[71]

(LiP, EC 1.11.1.14). For instance, UV-pretreated PE films were broken down by a copper-dependent laccase from *R. rubber* strain C208. Furthermore, the amorphous portion of PE films is oxidatively cleaved by laccase, which forms readily accessible carbonyl groups and significantly reduces the weight of the film. After using laccase from Trametes versicolor in the presence of 1-hydroxybenzotriazole as a mediator, Fujisawa et al. [21] demonstrated a decrease in the molecular weight of a PE membrane. UV-treated PE resulted in increased laccase and MnP secretion from B. cereus. On the other hand, MnP and a partly purified laccase from *P. simplicissimum* resulted in negligible weight loss. Heat-treated PE breakdown is aided by LiP activity in Streptomyces supernatants. Extracellular peroxidases are essential for the 70% breakdown of pre-oxidized high-molecular-weight PE that *P. chrysosporium* MTCC-787 demonstrated in 15 days.

3.3.1 Fungal Enzymes

Fungi are members of ten genera in the *Ascomycota* and *Mucor*, and they destroy polyethene faster than bacteria. This discrepancy is because fungus can cling to hydrophobic polymer surfaces. When PP and PE are the sole carbon sources, *Aspergillus, Fusarium oxysporum, and Penicillium* can be cultured for three months to examine their degradation capacity. The research also establishes the presence of biofilms and identifies surface changes in PP and PE. The incubated fungi are also investigated, revealing that they can persist for three months without additional carbon sources. Cutinases, esterases, lipases, laccases, peroxidases, proteases, and ureases from bacterial and fungal species can degrade PE, PET, and PP [72–74]. Fungal cellulase systems detected cellulose depolymerization-free enzymes acting directly on solid polymeric substrates, with the final step converting monomeric elements (for example, cellobiose hydrolysis to glucose) [57]. Fungal enzymes, particularly depolymerase, can break various polymers, including PET and PE. The dispersion and penetrative ability of fungal hyphae play an essential role in their initial colonization before depolymerization, as do their enzyme capabilities to increase hyphal adhesion to hydrophobic substrates. Laccases from actinomycetes, *Rhodococcus ruber*, and fungi, such as *Aspergillus flavus* and *Pleurotus ostreatus*, have all shown considerable degradation of PE. It could be caused by the oxidation of the PE hydrocarbon backbone [5]. Furthermore, in the biodegradation of typical low-density plastic LDPE, other fungal species with significant plastic degrading properties include *Fusarium solani, Alternaria solani, Aspergillus fumigatus, Spicaria spp., Geomyces pannorum, Phoma sp., Penicillium spp.*, etc. [58]. Bacteria and fungi, particularly *F. oxysporum and Aspergillus fumigates*, are the most common microbes that degrade polyethene materials. Cultivating non-degradable plastic waste PE with *F. oxysporum* strains improves tolerance to high-concentration PE, allowing for essential oxidation events and changes in PE film shape. *Aspergillus fumigates* produce biofilms to break down LDPE, obtaining the highest experimental efficiency in fungal LDPE degradation [59, 60]. Efficient bacteria like *Ideonella sakaiensis* 201-F or their enzymes, which include PETase and MHETase, destroy certain hydrolyzable polymers, including PET. Other investigations used hydrolytic enzymes, such as esterases and lipases, to degrade PET. PETase selectively hydrolyzes PET, converting it into the monomers ethylene glycol (EG) and terephthalic acid (TPA). PETase and MHETase collaborate to convert PET into these monomers, earning recognition for reducing plastic waste. PETase breaks PET stability by catalyzing ester linkages, releasing dimer BHET and monomer MHET, which MHETase then processes to form TPA and EG [74–79].

Because of their degrading solid ability and unique metabolic type, synergistic enzymes released by numerous microorganisms and bacteria with a two-enzyme system are the focus of research on the biodegradation of pollutants both internally and externally. Similarly, developing multienzyme systems for depolymerizing plastic waste could be a promising research field. For example, the *I. sakaiensis* strain contains a dual enzyme system composed of PETase and MHETase, which has evolved the ability to use crystalline polyester substrates [57]. Scientists have used genetic engineering to enhance further enzymes to overcome natural evolution. It must be purposefully modified to make the enzyme function more efficiently than in its native state [76–80].

An improved PET hydrolase depolymerizes PET into monomers with approximately 90% efficiency in 10 hours. The enzyme from the *I. sakaiensis* 201-F6 strain outperforms all others, efficiently disintegrating PET films in weeks at temperatures around 30 °C. Enzyme stability is linked to the enzyme–substrate interaction. Standard commercial methods, such as enzyme immobilization, can inhibit optimal PETase interactions, resulting in lower PET breakdown. Various cutinase enzymes capable of PET depolymerization have also been found in studies, showing the natural abundance of esterase enzymes. *Fusarium oxysporum*, an ascomycete, can release keratinase, which depolymerizes PET.

Similarly, the cutinase LC-cutinase, which can degrade PET and PCL, was derived from a leaf fosmid library and expressed in *Escherichia coli*. Bacterial and enzyme transformation and cloning have the potential to significantly increase plastic breakdown. Recent studies have shown that the two cutinases, LC-cutinase and *Thermobifida fusca* cutinase, achieved higher activity in the experiment, with the latter having higher residual activity after 40 hours. Thus, LC-cutinase has marginally lower stability than *T. fusca* cutinase [65]. In the presence of surfactants and organic solvents, the *T. fusca* cutinases displayed various benefits, including adaptive hydrolytic activity, good surfactant tolerance, superior stability in organic solvents, and thermostability.

According to the research report, the hydrolytic activity of LC-cutinase degrades PET. It exerts and improves particular skills and can degrade PET into TPA and EG [72]. Understanding the protein architectures of these enzymes allows us to improve their catalytic efficacy in degrading various polymers, including PET. Microorganisms evolved and discovered suitable enzymes for mutation, eventually selecting mutant keratinase to degrade PET rapidly [36].

3.3.2 Enzymes Produced by Microorganisms in the Extreme Environments

Halophiles and psychrophiles are extreme environmental microorganisms with plastic degradation capabilities. Bacteria in specific harsh settings can damage synthetic polymers. Extreme ecological conditions define plastic-contaminated areas, such as low or high temperatures, acidic or alkaline pH, high salt concentrations, and high pressure. Thermophilic and halophilic enzymes have a longer life cycle, allowing them to be stored at room temperature without losing significant enzymatic activity [74]. Hence, as a source of plastic-degrading enzymes and microorganisms, it is possible to research thermophile and extremophile microbiomes to exploit their biodegrading potential. At 65 °C, the most thermally stable leaf-branching compost cutinase (LCC) demonstrated the highest PET depolymerization rates [3]. Several thermophiles have shown a strong propensity for polymer degradation at elevated temperatures, akin to high-temperature plastic degradation agents. Bacteria can create many highly active enzymes, increasing substrate bioavailability and solubility [73]. As reported for the first time, *Chelatococcus sp.* E1 isolated from a compost sample could degrade PE at 60 °C. As thermophilic strains, PE co-cultured with *Chelatococcus sp.* E1 shifted the molecular weight distribution to the lower molecular weight side, increasing the biodegradability of HDPE and LDPE. One of society's most pressing issues could be addressed by producing extremozymes and encouraging the growth of extremophiles in harsh conditions.

3.4 Advanced Approaches

In the preceding section, the thrust was laid on the existing technologies available regarding plastic biodegradation via different mechanisms utilizing microbial diversity. However, this section will focus on specific advanced approaches in plastic biodegradation.

3.4.1 Bioplastics

Bioplastics are broadly defined as plastic materials manufactured from biodegradable and bio-based polymers. According to IUPAC, a bio-based polymer is obtained from biological sources such as plants, animals, and marine materials. Bio-based polymers are classified into three types: natural biopolymers (e.g., cellulose, starch), bio-derived polymers (e.g., cellulose acetate), and synthetic polymers made from renewable feedstock. Confusion surrounds the term "bioplastics" because "biodegradable" and "bio-based" define separate traits, and bioplastics can have one or both. Biodegradable plastics can be helpful when recycling is complex, and plastic waste may end up in the environment. However, completely replacing traditional plastics with biodegradable alternatives is neither a feasible nor preferable solution to plastic pollution [81–90].

3.4.2 Pretreatments and Effects of Additives for Enhancing Biodegradability of PE

I. Thermo-UV pretreatment

Thermo-UV pretreatment is used to cause partial photolysis in the PE layer and to emulate natural weathering. UV irradiation and heat treatments cause oxidation areas in the plastic's polymer chain, increasing its susceptibility to microbes. This procedure lowers the plastic's hydrophobicity and encourages the growth of microbial biofilms. Efforts to improve plastic biodegradation have investigated several pretreatments [17, 18]. According to studies, physicochemical pretreatments such as UV, thermo-oxidative, and chemical treatments can induce surface oxidation, resulting in the formation of carbonyl, carboxyl, and ester functional groups [19, 20], reducing surface hydrophobicity and promoting microbial biofilm formation, thereby improving biodegradation efficiency [22, 23].

Gamma irradiation and heat treatment can accelerate the breakdown of some polymers. QLAB, Homestead, and FL used a QUV-accelerated weathering tester to treat PE samples. PE was treated to UV and humidity cycles: five daily cycles of UV exposure (four of 4 hours each and one of 3 hours at 70°C) with 1-hour intervals at 50°C. Rhodococcus ruber (C208) tested the treated PE film for biodegradation and biofilm development. The FTIR examination of UV-treated PE revealed that carbonyl residues on the photo-oxidized surface are critical for beginning biodegradation [90–95].

Another study examined the *Bacillus amyloliquefaciens* strain isolated from composite plastic to see how thermo-irradiation pretreatment affected degradation. LDPE and LLDPE films were subjected to gamma rays and thermal treatment at 150 and 90 °C for 7 days, respectively. LDPE films were also subjected to UV irradiation in an oven at 60 °C for 7 days [79]. After 40–60 days of incubation with the isolated bacterium, the dry weight of pretreated LLDPE fell by 1.1 ± 0.3 to $3.2 \pm 1.3\%$. The FTIR spectra showed a flattening of the carbonyl band (1300-1100 cm^{-1}), indicating biodeterioration. Electro-spray ionization-mass spectrometry (ESI-MS) research revealed that 3-hydroxybutyrate oligomers were released exclusively in pretreated LLDPE, not native LLDPE [93–98]. These oligomers disappeared after incubation with *Bacillus amyloliquefaciens*, showing that low molecular LLDPE fractions were metabolized.

ii. Treatment with pro-oxidants
Stabilizers are added to commercially made PE films to prevent oxidation and increase life. Pro-oxidant additions, such as metal ions (iron, manganese, titanium, and cobalt), can boost photo- and thermo-oxidation, resulting in polymer chain cleavage. A study on LDPE with manganese stearate, followed by UV exposure and treatment with Aspergillus oryzae, found a 62% reduction in elongation and a 51% increase in tensile strength. FTIR measurement demonstrated increased carbonyl and carboxylic groups following pro-oxidant treatment, supporting their significance in PE biodegradation. Novotný et al. [101] identified the existence of an aldehyde. Ketone, ether, or ester groups were generated after gamma-irradiating cum heat-treated PE and decomposed more efficiently than the untreated sample. Taghavi et al. [102] discovered that UV pretreatment of PE and PS causes higher roughness, poorer hydrophobicity, increased biofilm colonization, and a more significant physical and molecular weight reduction than untreated samples. Chemical pretreatment of plastic effectively hydrolyzes the material and slows its breakdown. However, chemical pretreatments generate environmental problems because of the chemicals used, the cost, and the recovery of chemicals after treatment [24].

iii. Photocatalysis Using Titanium Dioxide (TiO$_2$)

TiO$_2$ is an environmentally friendly photocatalyst that absorbs ultraviolet light. As a result, polymer sheets containing TiO$_2$ absorb UV light efficiently. TiO$_2$-mediated photocatalysis involves the absorption of photons with appropriate energy, which generates electrons and holes that encourage the creation of free radicals, culminating in polymer oxidation and degradation. It has been examined the photocatalytic degradation of LDPE containing Titania nanoparticles after exposure to sunlight. After 200 hours of exposure to solar radiation, the composite PE film lost 68% of its weight, much higher than the rate of weight loss recorded after 400 hours in a previous study [32]. FTIR and SEM analyses indicated the presence of carbonyl. Another study found that sun irradiation of a PE film blended with copper phthalocyanine (CuPc) modified TiO$_2$ (TiO$_2$/CuPc) photocatalyst resulted in a substantial weight loss rate, rough surface texture, and an enormous amount of CO$_2$ created compared to the original PE film. Surface photovoltage spectroscopy (SPS) study indicated that CuPc increased TiO$_2$ charge separation. The increased breakdown of PE is due to reactive oxygen species produced on the surface of TiO$_2$ or TiO$_2$/CuPc particles. Fa et al. [32] recently made a TiO$_2$–FeSt$_3$ ferric stearate–polyethylene (TFPE) composite film and investigated photo-degradation under UV irradiation for 240 hours and thermo-degradation at 70 °C for 30 days. FTIR spectroscopy indicates the development of carbonyl and hydroxyl groups, which help biodegrade PE films.

Other research developments

Many research studies have studied PE-degrading bacteria using commercially available polymers that may have diverse chemical additives. The level of degradation was determined by measuring weight loss and functional group changes on the polymer surface using FTIR. However, it is difficult to determine if the weight loss and surface structure changes are caused by the breakdown of additives, which frequently account for a significant amount of the polymer. As a result, more significant adjustments are required to identify actual PE degradation and reduce the likelihood of artefacts caused by additive degradation [22, 86, 95–97]. In this context, a robust, reliable approach has been established to analyze the biodegradability of PE by quantifying CO$_2$ using gas chromatography due to bacterial degradation and respiration [90].

The soil bacterium *R. rhodochrous* was cultivated in a specified aqueous medium with PE as the sole carbon source, and CO$_2$ production was proportional to the mineralization of the supplied carbon source via bacterial respiration. At the stationary phase, there is no substantial difference in CO$_2$ release between cells cultivated without a carbon supply and with LDPE. This indicated that carbon bioavailability was restricted in LDPE bacterial growth, resulting in limited biodegradation. Furthermore, UV pretreatment's effect on LDPE's biodegradability was investigated by incubating *R. rhodochrous* with UV-pretreated and native LDPE for 35 days and measuring the amount of CO$_2$ emitted over time. UV-treated LDPE was three times more biodegradable than non-treated LDPE. In one work, 1320 oxidized oligomers from PE films were examined before and after biodegradation by *R. rhodochrous* using MS and NMR spectroscopy. After 240 days, the strain assimilated 95% of soluble oligomers. Longer molecules degraded faster, highlighting the role of extracellular chain cleavage and intracellular β-oxidation in PE biodegradation. Additional research assesses plastic biodegradability via carbon tracing to CO_2 and biomass.

3.5   Conclusion and Future Prospective

Until now, much PE biodegradation research has explored changes in physicochemical qualities and structural deterioration utilizing FTIR, DSC, XRD, SEM, AFM, etc. Predominantly measured weight loss and physicochemical alterations cannot demonstrate PE's actual biodegradation. There is a need to offer specific and credible proof for PE biodegradation to reduce artefacts caused by additive degradation rather than PE. As a result, future research should focus on additive-free PE. Furthermore, 13C-polyethylene degradation is proposed to demonstrate the progressive synthesis of 13C-labelled metabolites, including CO_2 emissions, throughout an incubation period. Further research into the enzymatic degradation process will reveal the molecular pathway for efficient biodegradation of polyethene [93–100].

References

1. Khan, F.; Ahmed, W.; Najmi, A.; Younus, M. Managing plastic waste disposal by assessing consumers' recycling behavior: The case of a densely populated developing country. Environ. Sci. Pollut. Res. 2019, 26, 33054–33066.
2. Yao, Z.; Seong, H.J.; Jang, Y.S. Environmental toxicity and decomposition of polyethylene. Ecotoxicol. Environ. Saf. 2022, 242, 113933.
3. Wang, S.; Shi, W.; Huang, Z.; Zhou, N.; Xie, Y.; Tang, Y.; Hu, F.; Liu, G.; Zheng, H. Complete digestion/biodegradation of polystyrene microplastics by greater wax moth (Galleria mellonella) larvae: Direct in vivo evidence, gut microbiota independence, and potential metabolic pathways. J. Hazard. Mater. 2022, 423, 127213.
4. Yang, S.-S.; Ding, M.-Q.; He, L.; Zhang, C.-H.; Li, Q.-X.; Xing, D.-F.; Cao, G.-L.; Zhao, L.; Ding, J.; Ren, N.-Q.; et al. Biodegradation of polypropylene by yellow mealworms (Tenebrio molitor) and superworms (Zophobas atratus) via gut-microbe-dependent depolymerization. Sci. Total Environ. 2021, 756, 144087.
5. Awasthi, S.K.; Kumar, M.; Kumar, V.; Sarsaiya, S.; Anerao, P.; Ghosh, P.; Singh, L.; Liu, H.; Zhang, Z.; Awasthi, M.K. A comprehensive review on recent advancements in biodegradation and sustainable management of biopolymers. Environ. Pollut. 2022, 307, 119600.
6. Rafey, A.; Pal, K.; Bohre, A.; Modak, A.; Pant, K.K. A State-of-the-Art Review on the Technological Advancements for the Sustainable Management of Plastic Waste in Consort with the Generation of Energy and Value-Added Chemicals. Catalysts 2023, 13, 420.
7. Jambeck, J.R.; Geyer, R.; Wilcox, C.; Siegler, T.R.; Perryman, M.; Andrady, A.; Narayan, R.; Law, K.L. Plastic waste inputs from land into the ocean. Science 2015, 347, 768–771.

8. Asiandu, A.P.; Wahyudi, A.; Sari, S.W. Aquatic plastics waste biodegradation using plastic degrading microbes. J. Microbiol. Biotechnol. Food Sci. 2022, 11, e3724.

9. Chea, J.D.; Yenkie, K.M.; Stanzione, J.F., III; Ruiz-Mercado, G.J. A generic scenario analysis of end-of-life plastic management: Chemical additives. J. Hazard. Mater. 2023, 441, 129902.

10. Lin, Z.; Jin, T.; Zou, T.; Xu, L.; Xi, B.; Xu, D.; He, J.; Xiong, L.; Tang, C.; Peng, J.; et al. Current progress on plastic/microplastic degradation: Fact influences and mechanism. Environ. Pollut. 2022, 304, 119159.

11. Gewert, B.; Plassmann, M.M.; MacLeod, M. Pathways for degradation of plastic polymers floating in the marine environment. Environ. Sci. Process. Impacts 2015, 17, 1513–1521. Int. J. Mol. Sci. 2024, 25, 593 18 of 25.

12. Ferreira, G.V.B.; Justino, A.K.S.; Eduardo, L.N.; Schmidt, N.; Martins, J.R.; Menard, F.; Fauvelle, V.; Mincarone, M.M.; Lucena- Fredou, F. Influencing factors for microplastic intake in abundant deep-sea lanternfishes (Myctophidae). Sci. Total Environ. 2023, 867, 161478.

13. Pereira, R.; Rodrigues, S.M.; Silva, D.; Freitas, V.; Almeida, C.M.R.; Ramos, S. Microplastic contamination in large migratory fishes collected in the open Atlantic Ocean. Mar. Pollut. Bull. 2023, 186, 114454.

14. Lee, H.C.; Khan, M.M.; Yusli, A.A.; Jaya, N.A.; Marshall, D.J. Microplastic accumulation in oysters along a Bornean coastline (Brunei, South China Sea): Insights into local sources and sinks. Mar. Pollut. Bull. 2022, 177, 113478.

15. Song, J.A.; Choi, C.Y.; Park, H.-S. Exposure of bay scallop Argopecten irradians to micropolystyrene: Bioaccumulation and toxicity. Comp. Biochem. Physiol. Part C Toxicol. Pharmacol. 2020, 236, 108801.

16. Jiang, Q.; Chen, X.; Jiang, H.; Wang, M.; Zhang, T.; Zhang, W. Effects of Acute Exposure to Polystyrene Nanoplastics on the Channel Catfish Larvae: Insights From Energy Metabolism and Transcriptomic Analysis. Front. Physiol. 2022, 13, 923278.

17. Schmidt, A.; Brito, W.A.d.S.; Singer, D.; Muehl, M.; Berner, J.; Saadati, F.; Wolff, C.; Miebach, L.; Wende, K.; Bekeschus, S. Short- and long-term polystyrene nano- and microplastic exposure promotes oxidative stress and divergently affects skin cell architecture and Wnt/beta-catenin signaling. Part. Fibre Toxicol. 2023, 20, 3.

18. Lu, Y.; Huang, R.; Wang, J.; Wang, L.; Zhang, W. Effects of Polyester Microfibers on the Growth and Toxicity Production of Bloom-Forming Cyanobacterium Microcystis aeruginosa. Water 2022, 14, 2422.

19. Wu, D.; Wang, T.; Wang, J.; Jiang, L.; Yin, Y.; Guo, H. Size-dependent toxic effects of polystyrene microplastic exposure on Microcystis aeruginosa growth and microcystin production. Sci. Total Environ. 2021, 761, 143265.

20. Chen, Y.; Ling, Y.; Li, X.; Hu, J.; Cao, C.; He, D. Size-dependent cellular internalization and effects of polystyrene microplastics in microalgae P. helgolandica var. tsingtaoensis and S. quadricauda. J. Hazard. Mater. 2020, 399, 123092.

21. Fujisawa M, Hirai H, Nishida T (2001) Degradation of polyethylene and nylon-66 by the laccase-mediator system. J Polym Environ 9:103–108

22. Fauser, P.; Vorkamp, K.; Strand, J. Residual additives in marine microplastics and their risk assessment—A critical review. Mar. . Bull. 2022, 177, 113467.

23. Song, Y.K.; Hong, S.H.; Jang, M.; Han, G.M.; Jung, S.W.; Shim, W.J. Combined Effects of UV Exposure Duration and Mechanical Abrasion on Microplastic Fragmentation by Polymer Type. Environ. Sci. Technol. 2018, 52, 3831–3832.

24. Barrick, A.; Champeau, O.; Chatel, A.; Manier, N.; Northcott, G.; Tremblay, L.A. Plastic additives: Challenges in ecotox hazard assessment. PeerJ 2021, 9, e11300. [CrossRef].

25. Sanchez, C. Microbial capability for the degradation of chemical additives present in petroleum-based plastic products: A review on current status and perspectives. J. Hazard. Mater. 2021, 402, 123534.

26. Ren, L.; Lin, Z.; Liu, H.; Hu, H. Bacteria-mediated phthalic acid esters degradation and related molecular mechanisms. Appl. Microbiol. Biotechnol. 2018, 102, 1085–1096. [CrossRef].

27. Lumio, R.T.; Tan, M.A.; Magpantay, H.D. Biotechnology-based microbial degradation of plastic additives. 3 Biotech 2021, 11, 350.

28. Staples, C.A.; Peterson, D.R.; Parkerton, T.F.; Adams, W.J. The environmental fate of phthalate esters: A literature review. Chemosphere 1997, 35, 667–749.

29. Ramzi, A.; Gireeshkumar, T.R.; Rahman, K.H.; Balachandran, K.K.; Shameem, K.; Chacko, J.; Chandramohanakumar, N. Phthalic acid esters—A grave ecological hazard in Cochin estuary, India. Mar. Pollut. Bull. 2020, 152, 110899.

30. Pan, X.; Liu, A.; Zheng, M.; Liu, J.; Du, M.;Wang, L. Determination and environmental risk assessment of organophosphorus flame retardants in sediments of the South China Sea. Environ. Sci. Pollut. Res. 2022, 29, 70542–70551.

31. Wang, S.; Sun, Z.; Ren, C.; Li, F.; Xu, Y.; Wu, H.; Ji, C. Time- and dose-dependent detoxification and reproductive endocrine disruption induced by tetrabromobisphenol A (TBBPA) in mussel Mytilus galloprovincialis. Mar. Environ. Res. 2023, 183, 105839.

32. Chackal, R.; Eng, T.; Rodrigues, E.M.; Matthews, S.; Page-Lariviere, F.; Avery-Gomm, S.; Xu, E.G.; Tufenkji, N.; Hemmer, E.; Mennigen, J.A. Metabolic Consequences of Developmental Exposure to Polystyrene Nanoplastics, the Flame Retardant BDE-47 and Their Combination in Zebrafish. Front. Pharmacol. 2022, 13, 822111. [CrossRef].

33. Chen, X.; Chen, C.-E.; Guo, X.; Sweetman, A.J. Sorption and desorption of bisphenols on commercial plastics and the effect of UV aging. Chemosphere 2023, 310, 136867. [CrossRef].

34. Andelic, I.; Roje-Busatto, R.; Ujevic, I.; Vuletic, N.; Matijevic, S. Distribution of Bisphenol A in Sediment and Suspended Matter and Its Possible Impact on Marine Life in Kastela Bay, Adriatic Sea, Croatia. J. Mar. Sci. Eng. 2020, 8, 480. [CrossRef].

35. Huang, Q.; Liu, Y.; Chen, Y.; Fang, C.; Chi, Y.; Zhu, H.; Lin, Y.; Ye, G.; Dong, S. New insights into the metabolism and toxicity of bisphenol A on marine fish under long-term exposure. Environ. Pollut. 2018, 242, 914–921.

36. Harshvardhan, K.; Jha, B. Biodegradation of low-density polyethylene by marine bacteria from pelagic waters, Arabian Sea, India. Mar. Pollut. Bull. 2013, 77, 100–106.

37. Oliveira, J.; Belchior, A.; da Silva, V.D.; Rotter, A.; Petrovski, Z.; Almeida, P.L.; Lourenco, N.D.; Gaudencio, S.P. Marine Environmental Plastic Pollution: Mitigation by Microorganism Degradation and Recycling Valorization. Front. Mar. Sci. 2020, 7, 567126

38. Lange, J.-P. Managing Plastic Waste-Sorting, Recycling, Disposal, and Product Redesign. ACS Sustain. Chem. Eng. 2021, 9, 15722–15738.

39. Li,W.; Zhao,W.; Zhu, H.; Li, Z.-J.;Wang,W. State of the art in the photochemical degradation of (micro)plastics: From fundamental principles to catalysts and applications. J. Mater. Chem. A 2023, 11, 2503–2527.

40. Ncube, L.K.; Ude, A.U.; Ogunmuyiwa, E.N.; Zulkifli, R.; Beas, I.N. An Overview of Plastic Waste Generation and Management in Food Packaging Industries. Recycling 2021, 6, 12

41. Jaiswal, S.; Sharma, B.; Shukla, P. Integrated approaches in microbial degradation of plastics. Environ. Technol. Innov. 2020, 17, 100567.

42. Pivato, A.F.; Miranda, G.M.; Prichula, J.; Lima, J.E.A.; Ligabue, R.A.; Seixas, A.; Trentin, D.S. Hydrocarbon-based plastics: Progress and perspectives on consumption and biodegradation by insect larvae. Chemosphere 2022, 293, 133600.

43. Yoshida, S.; Hiraga, K.; Takehana, T.; Taniguchi, I.; Yamaji, H.; Maeda, Y.; Toyohara, K.; Miyamoto, K.; Kimura, Y.; Oda, K. A bacterium that degrades and assimilates poly(ethylene terephthalate). Science 2016, 351, 1196–1199.
44. Yang, S.S.; Wu, W.M.; Pang, J.-W.; He, L.; Ding, M.-Q.; Li, M.-X.; Zhao, Y.-L.; Sun, H.J.; Xing, D.-F.; Ren, N.-Q.; et al. Bibliometric analysis of publications on biodegradation of plastics: Explosively emerging research over 70 years. J. Cleaner Prod. 2023, 428, 139423.
45. Anand, U.; Dey, S.; Bontempi, E.; Ducoli, S.; Vethaak, A.D.; Dey, A.; Federici, S. Biotechnological methods to remove microplastics: A review. Environ. Chem. Lett. 2023, 21, 1787–1810.
46. Bacha, A.-U.-R.; Nabi, I.; Zaheer, M.; Jin, W.; Yang, L. Biodegradation of macro- and microplastics in environment: A review on mechanism, toxicity, and future perspectives. Sci. Total Environ. 2023, 858, 160108.
47. Cacciari, I.; Quatrini, P.; Zirletta, G.; Mincione, E.; Vinciguerra, V.; Lupattelli, P.; Sermanni, G.G. Isotactic polypropylene biodegradation by a microbial community—Physicochemical characterization of metabolites produced. Appl. Environ. Microbiol. 1993, 59, 3695–3700.
48. Lee, H.M.; Kim, H.R.; Jeon, E.; Yu, H.C.; Lee, S.; Li, J.; Kim, D.-H. Evaluation of the Biodegradation Efficiency of Four Various Types of Plastics by Pseudomonas aeruginosa Isolated from the Gut Extract of Superworms. Microorganisms 2020, 8, 1341.
49. Gupta, K.K.; Devi, D. Characteristics investigation on biofilm formation and biodegradation activities of Pseudomonas aeruginosa strain ISJ14 colonizing low density polyethylene (LDPE) surface. Heliyon 2020, 6, e04398.
50. Giacomucci, L.; Raddadi, N.; Soccio, M.; Lotti, N.; Fava, F. Polyvinyl chloride biodegradation by Pseudomonas citronellolis and Bacillus flexus. New Biotechnol. 2019, 52, 35–41.
51. Fontanazza, S.; Restuccia, A.; Mauromicale, G.; Scavo, A.; Abbate, C. Pseudomonas putida Isolation and Quantification by Real-Time PCR in Agricultural Soil Biodegradable Mulching. Agriculture 2021, 11, 782.
52. Miloloza, M.; Ukic, S.; Cvetnic, M.; Bolanca, T.; Grgic, D.K. Optimization of Polystyrene Biodegradation by Bacillus cereus and Pseudomonas alcaligenes Using Full Factorial Design. Polymers 2022, 14, 4299.
53. Wrobel, M.; Szymanska, S.; Kowalkowski, T.; Hrynkiewicz, K. Selection of microorganisms capable of polyethylene (PE) and polypropylene (PP) degradation. Microbiol. Res. 2023, 267, 127251.
54. Lacerda, A.L.d.F.; Taylor, J.D.; Rodrigues, L.D.S.; Kessler, F.; Secchi, E.; Proietti, M.C. Floating plastics and their associated biota in the Western South Atlantic. Sci. Total Environ. 2022, 805, 150186.
55. Dong, X.; Zhu, L.; Jiang, P.; Wang, X.; Liu, K.; Li, C.; Li, D. Seasonal biofilm formation on floating microplastics in coastal waters of intensified marinculture area. Mar. Pollut. Bull. 2021, 171, 112914.
56. Bollinger, A.; Thies, S.; Knieps-Gruenhagen, E.; Gertzen, C.; Kobus, S.; Hoeppner, A.; Ferrer, M.; Gohlke, H.; Smits, S.H.J.; Jaeger, K.-E. A Novel Polyester Hydrolase From the Marine Bacterium Pseudomonas aestusnigri—Structural and Functional Insights. Front. Microbiol. 2020, 11, 114.
57. Gomila, M.; Mulet, M.; Lalucat, J.; Garcia-Valdes, E. Draft Genome Sequence of the Marine Bacterium Pseudomonas aestusnigri VGXO14T. Genome Announc. 2017, 5, e00765-17.
58. Oliveira, M.M.; Proenca, A.M.; Moreira-Silva, E.; de Castro, A.M.; dos Santos, F.M.; Marconatto, L.; Medina-Silva, R. Biofilms of Pseudomonas and Lysinibacillus Marine Strains on High-Density Polyethylene. Microb. Ecol. 2021, 81, 833–846.
59. Ghatge, S., Yang, Y., Ahn, JH. et al. Biodegradation of polyethylene: a brief review. Appl Biol Chem 63, 27 (2020). https://doi.org/10.1186/s13765-020-00511-3

60. Skariyachan, S.; Megha, M.; Kini, M.N.; Mukund, K.M.; Rizvi, A.; Vasist, K. Selection and screening of microbial consortia for efficient and ecofriendly degradation of plastic garbage collected from urban and rural areas of Bangalore, India. Environ. Monit. Assess. 2015, 187, 4174.

61. Auta, H.S.; Emenike, C.U.; Fauziah, S.H. Screening of Bacillus strains isolated from mangrove ecosystems in Peninsular Malaysia for microplastic degradation. Environ. Pollut. 2017, 231, 1552–1559.

62. Zheng Y, Yanful EK, Bassi AS (2005) A review of plastic waste biodegradation. Crit Rev Biotechnol 25:243–250

63. Yoon MG, Jeon HJ, Kim MN (2012) Biodegradation of polyethylene by a soil bacterium and AlkB cloned recombinant cell. J Bioremed Biodegrad 3:145

64. Khandare, S.D.; Agrawal, D.; Mehru, N.; Chaudhary, D.R. Marine bacterial based enzymatic degradation of low-density polyethylene (LDPE) plastic. J. Environ. Chem. Eng. 2022, 10, 107437

65. Kumar, A.G.; Hinduja, M.; Sujitha, K.; Rajan, N.N.; Dharani, G. Biodegradation of polystyrene by deep-sea Bacillus paralicheniformi G1 and genome analysis. Sci. Total Environ. 2021, 774, 145002.

66. Maity, W.; Maity, S.; Bera, S.; Roy, A. Emerging Roles of PETase and MHETase in the Biodegradation of Plastic Wastes. Appl. Biochem. Biotechnol. 2021, 193, 2699–2716

67. Carmen, S. Microbial capability for the degradation of chemical additives present in petroleum-based plastic products: A review on current status and perspectives. J. Hazard. Mater. 2021, 402, 123534.

68. Das, R.; Liang, Z.; Li, G.; Mai, B.; An, T. Genome sequence of a spore-laccase forming, BPA-degrading Bacillus sp. GZB isolated from an electronic-waste recycling site reveals insights into BPA degradation pathways. Arch. Microbiol. 2019, 201, 623–638.

69. Feng, S.; Yue, Y.; Chen, J.; Zhou, J.; Li, Y.; Zhang, Q. Biodegradation mechanism of poly-caprolactone by a novel esterase MGS0156: A QM/MM approach. Environ. Sci. Process. Impacts 2020, 22, 2332–2344.

70. Sulaiman, S.; Yamato, S.; Kanaya, E.; Kim, J.J.; Koga, Y.; Takano, K.; Kanaya, S. Isolation of a novel cutinase homolog with polyethylene terephthalate-degrading activity from leaf-branch compost by using a metagenomic approach. Appl. Environ. Microbiol. 2012, 78, 1556–1562

71. Atanasova, N.; Stoitsova, S.; Paunova-Krasteva, T.; Kambourova, M. Plastic Degradation by Extremophilic Bacteria. Int. J. Mol. Sci. 2021, 22, 5610

72. Kopecka, R.; Kubinova, I.; Sovova, K.; Mravcova, L.; Vitez, T.; Vitezova, M. Microbial degradation of virgin polyethylene by bacteria isolated from a landfill site. SN Appl. Sci. 2022, 4, 302.

73. Gupta, K.K.; Sharma, K.K.; Chandra, H. Utilization of Bacillus cereus strain CGK5 associated with cow feces in the degradation of commercially available high-density polyethylene (HDPE). Arch. Microbiol. 2023, 205, 101.

74. Das, M.P.; Kumar, S. An approach to low-density polyethylene biodegradation by Bacillus amyloliquefaciens. 3 Biotech 2015, 5, 81–86.

75. Dede, B.; Priest, T.; Bach, W.; Walter, M.; Amann, R.; Meyerdierks, A. High abundance of hydrocarbon-degrading Alcanivorax in plumes of hydrothermally active volcanoes in the South Pacific Ocean. ISME J. 2023, 17, 600–610.

76. Parray, J.A., Haghi, A.K., Meraj, G. (eds) 2024 IoT-Based Models for Sustainable Environmental Management. Lecture Notes on Data Engineering and Communications Technologies, vol 227. Springer, Cham. ISBN-978-3-031-74373-3.

77. Parray, J.A., Shameem N., Haghi, A.K., (eds) 2024 Management of Waste to Control Environmental Pollutions: Sustainability and Economic Feasibility Springer, Cham. ISBN: 978-3-031-80844-9.

78. Cappello, S.; Caruso, G.; Bergami, E.; Macri, A.; Venuti, V.; Majolino, D.; Corsi, I. New insights into the structure and function of the prokaryotic communities colonizing plastic debris collected in King George Island (Antarctica): Preliminary observations from two plastic fragments. J. Hazard. Mater. 2021, 414, 125586.

79. Zhao, S.; Liu, R.;Wang, J.; Lv, S.; Zhang, B.; Dong, C.; Shao, Z. Biodegradation of polyethylene terephthalate (PET) by diverse marine bacteria in deep-sea sediments. Environ. Microbiol. 2023, 25, 2719–2731.

80. Zadjelovic, V.; Erni-Cassola, G.; Obrador-Viel, T.; Lester, D.; Eley, Y.; Gibson, M.I.; Dorador, C.; Golyshin, P.N.; Black, S.;Wellington, E.M.H.; et al. A mechanistic understanding of polyethylene biodegradation by the marine bacterium Alcanivorax. J. Hazard. Mater. 2022, 436, 129278.

81. Delacuvellerie, A.; Cyriaque, V.; Gobert, S.; Benali, S.;Wattiez, R. The plastisphere in marine ecosystem hosts potential specific microbial degraders including Alcanivorax borkumensis as a key player for the low-density polyethylene degradation. J. Hazard Mater. 2019, 380, 120899.

82. Parray, J.A., Yaseen Mir, M., Haghi, A.K. (2024). Enzymatic Degradation of Synthetic Plastics: New Insights. In: Enzymes in Environmental Management. SpringerBriefs in Environmental Science. Springer, Cham. https://doi.org/10.1007/978-3-031-74874-5_2.

83. Sekiguchi, T.; Saika, A.; Nomura, K.; Watanabe, T.; Watanabe, T.; Fujimoto, Y.; Enoki, M.; Sato, T.; Kato, C.; Kanehiro, H. Biodegradation of aliphatic polyesters soaked in deep seawaters and isolation of poly(epsilon-caprolactone)-degrading bacteria. Polym. Degrad. Stab. 2011, 96, 1397–1403.

84. Liu, R.; Zhao, S.; Zhang, B.; Li, G.; Fu, X.; Yan, P.; Shao, Z. Biodegradation of polystyrene (PS) by marine bacteria in mangrove ecosystem. J. Hazard. Mater. 2023, 442, 130056.

85. Qi, X.; Ren, Y.;Wang, X. New advances in the biodegradation of Poly(lactic) acid. Int. Biodeterior. Biodegrad. 2017, 117, 215–223.

86. Santo M, Weitsman R, Sivan A (2013) The role of the copper-binding enzyme - laccase - in the biodegradation of polyethylene by the actinomycete Rhodococcus ruber. Int Biodeterior Biodegradation 84:204–210

87. Satlewal A, Soni R, Zaidi M, Shouche Y, Goel R (2008) Comparative biodegradation of HDPE and LDPE using an indigenously developed microbial consortium. J Microbiol Biotechnol 18:477–482

88. Secchi ER, Zarzur S (1999) Plastic debris ingested by a Blainville's beaked whale, Mesoplodon densirostris, washed ashore in Brazil. Aquat Mamm 25:21–24

89. Sen SK, Raut S (2015) Microbial degradation of low density polyethylene (LDPE): a review. J Environ Chem Eng 3:462–473

90. Seneviratne G, Tennakoon N, Weerasekara M, Nandasena K (2006) Polyethylene biodegradation by a developed Penicillium-Bacillus biofilm. Curr Sci 90:20–21

91. Shah AA, Hasan F, Hameed A, Ahmed S (2008) Biological degradation of plastics: a comprehensive review. Biotechnol Adv 26:246–265

92. Shimao M (2001) Biodegradation of plastics. Curr Opin Biotechnol 12:242–247

93. Sivan A (2011) New perspectives in plastic biodegradation. Curr Opin Biotech 22:422–426

94. Sivan A, Szanto M, Pavlov V (2006) Biofilm development of the polyethylene degrading bacterium Rhodococcus ruber. Appl Microbiol Biotechnol 72:346–352

95. Sowmya HV, Ramalingappa Krishnappa M, Thippeswamy B (2014) Biodegradation of polyethylene by Bacillus cereus. Adv Polym Sci Technol Int J 4:28–32

96. Sowmya HV, Ramalingappa Krishnappa M, Thippeswamy B (2015) Degradation of polyethylene by Penicillium simplicissimum isolated from local dumpsite of shivamogga district. Environ Dev Sustain 17:731–745

97. Spear LB, Ainley DG, Ribic CA (1995) Incidence of plastic in seabirds from the tropical pacific, 1984–1991: relation with distribution of species, sex, age, season, year and body weight. Mar Environ Res 40:123–146

98. Sudhakar M, Doble M, Murthy PS, Venkatesan R (2008) Marine microbe-mediated biodegradation of low- and high-density polyethylenes. Int Biodeterior Biodegradation 61:203–213

99. Parray, J.A., Shameem N., Haghi, A.K., (eds) 2025 Sustainable Urban Environment and Waste Management: Theory and practices Springer Nature Singapore Pte Ltd. ISBN: 978-981-96-1139-3

100. Oliveira, J.; Almeida, P.L.; Sobral, R.G.; Lourenco, N.D.; Gaudencio, S.P. Marine-Derived Actinomycetes: Biodegradation of Plastics and Formation of PHA Bioplastics—A Circular Bioeconomy Approach. Mar. Drugs 2022, 20, 760.

101 Novotný, Č.; Malachová, K.; Adamus, G.; Kwiecień, M.; Lotti, N.; Soccio, M.; Verney, V.; Fava. F. (2018) Deterioration of irradiation/high-temperature pretreated, linear low-density polyethylene (LLDPE) by Bacillus amyloliquefaciens Int. Biodeterior. Biodegrad. 132, pp. 259–267

102 Taghavi, N.; Singhal, N.; Zhuang, W.-Q.; Baroutian, S. (2021) Degradation of plastic waste using stimulated and naturally occurring microbial strains Chemosphere, 263, Article 127975

Biodegradation Potential of Microbes from Extreme Environments

4.1 Biodegradation of Plastics from Oceans

If people, governments, and industrial sectors take biodiversity into consideration when making decisions globally, anthropogenic consequences that threaten our environment, food supply, and the people who depend on that food source can be reduced. However, to guarantee sustained food security and the preservation of our biodiversity, we must preserve our current marine ecosystem and restore coastal areas. Globally, the yearly production of plastic has grown dramatically in recent years. The world's oceans are home to plastics, which have an adverse effect on marine life. The chemical makeup, nutrition availability, organisms, and time all affect how biodegradable plastics are [1]. Polyethylene terephthalate (PET), polyethylene (PE), including low and high-density PE, polyvinyl chloride (PVC), polypropylene (PP), polyurethane (PUR), polystyrene (PS), and expanded polystyrene (EPS) are the plastic polymers that are produced at the highest rates [2]. It can take decades or even centuries for plastics and organic debris to decompose in the water. Most plastics can decompose into macro- or microplastics and do not biodegrade rapidly. Plastic garbage is regarded as harmful to the ecosystem since it may be consumed by or entangled in marine life. Pollution from small, single-use plastics is happening and is having a big effect on the marine ecosystem. The biodegradation period for dissolved organic material and particulate organic material can range from days to months at depths as low as 40 m, months to years at depths as low as 200 m, and centuries at deeper depths. Approximately 76% of all bacteria worldwide are thought to be found in oceanic and coastal sediments. The deep-sea benthic boundary layer (BBL), the sediment–water interface (SWI), and the sea bottom in the northeastern Atlantic Ocean are areas with limited food supplies. Temperatures are low at 2 °C and pressures are high at 450 atm. Biogeochemical cycles (BC) are the processes by which microorganisms consume substances that are useful to them. Plastics in the water may impact the elements that make up the BC—carbon, phosphorus, sulphur, chromium, nitrogen, calcium, iron, silicate, and manganese [3]. Research revealed that cyanobacteria were more prevalent on plastic than on organic particles or non-particulate microorganisms in the surrounding marine environment. Additional variables including temperature, phosphorus, nitrogen, and oxygen levels might restrict bacterial development and the biodegradation of plastics in the ocean. Most organisms cannot access the inert dissolved N_2 gas (more than 95%) found in the marine environment. A tiny percentage of nitrogen in the seas is reactive nitrogen, which comprises accessible dissolved organic nitrogen, ammonia, and nitrate. The carbon/phosphorous ratio (median 163), nitrogen/phosphorous ratio (median 22), and carbon/nitrogen ratio (median 6) were all scattered across a large range of data in the study. Eighty-nine per cent of the samples included in this study were from the top 200 m of the ocean. According to Martiny et al. [4], the data is skewed towards the biogeochemical system that is found in the upper 200 m. Depending on the limiting nutrient, nitrogen/ phosphorus ratios in culture can range from 5:1 to 100:1 for algae and cyanobacteria. For three billion years, cyanobacteria dominated the ocean's phytoplankton. Throughout the billions of years of Earth's history, cyanobacteria thrived despite nutrient shortages because of their adaptability. According to [5], there is general agreement that nitrogen is limiting on short durations. In the marine environment, bacteria that live on plastics are more varied and flourishing than bacteria that live nearby. Ocean life may be impacted by BC's response to plastic pollution [3]. It can get as chilly as 2 °C in the high latitude polar seas. Temperatures near the equator can reach 36 °C, or top millimetres. According to the Argovis data, the temperature near the equator is 29 °C at a depth of 10 m and drops with increasing depth. The Arctic and Antarctic sites had the highest dissolved oxygen levels, while those closest to the equator had the lowest values. As depth grew, so did nitrate. The American Society for Testing Materials (ASTM), the International Organization for Standardization (ISO), and the U.S. National Oceanic and Atmospheric Administration (NOAA) have all published plastic methods for the analysis of synthetic materials, man-made materials, and/biodegradation.

For microplastics (less than 5 mm and greater than 0.3 mm), NOAA offers an analysis procedure. To digest organic stuff in a readable manner, the process calls for wet peroxide oxidation using a Fe (II) catalyst. This technique is used to identify PS, PVC, PP, and PE plastics [6]. Many other kinds of plastic and synthetic fibres that have been found in the marine environment would not be appropriate for this approach. Man-made materials, plastics, and synthetic materials are tested using ASTM test method D6691-17, Determination of Aerobic Biodegradation of Plastic Materials in the Marine Environment by a Defined Microbial Consortium or Natural Sea Water Inoculum. To determine the level of biodegradability at 30 °C, the technique uses a natural seawater sample or a homogenous inoculum supplemented with inorganic nutrients. There are approximately 635 mg/ L of ammonia and 307 mg/L of nitrate in the minimal marine solution. According to Bagheri et al. [7], the ocean's dissolved inorganic nitrogen content is roughly 5.8 umols/ L, or 0.1 mg/ammonium. The minimal marine solution has an ammonia content that is almost 6000 times greater than that of the ocean. According to the ASTM technique, the nitrate concentration in the lowest marine solution is roughly 110 times greater than what is found in the ocean. The Indian Ocean had zero nitrate levels close to the surface.

The aerobic conditions in the water serve as the basis for ASTM testing. There isn't much time spent close to the surface. Plastics produced of PE or PP for fishing nets and ropes may end up close to the ocean's surface. Biofilm growth causes other buoyant plastics to sink [3]. Using a safe respirometer, the ISO method 14,851:2019, Determination of the Ultimate Aerobic Biodegradability of Plastic Materials in an Aqueous Medium, calculates the amount of oxygen needed for biodegradation. Since activated sludge is only employed in fresh water, marine circumstances are not included in this procedure. Although test material may be delivered as films, bits, fragments, or formed items, the ISO standard prefers that it be used in powder form. The rate of biodegradation may be overestimated if the plastic is not in its natural state since a large surface area may allow more bacteria to break down the substance. Additionally, the guideline recommended using polymers free of additives like plasticizers. Once more, this would skew the results in favour of greater biodegradation than what is released into lakes, streams, and rivers. Since activated sludge supernatant is not prevalent in the natural environment, using it in the inoculum would skew the biodegradation time. The standard test medium contains 113 mg/L of phosphorus. Compared to fresh water, this is magnitudes higher. The phosphorous content must match that of the natural environment in order to extrapolate the findings to biodegradation in the natural setting. Time frame and lower temperatures may greatly lengthen the biodegradation period and provide the appearance that biodegradation is occurring naturally in fresh water. Phosphorus has been identified as the limiting nutrient in fresh water, while nitrogen is the limiting nutrient in marine environments. Biofilm respiration on plastic and other plastic-related compounds may affect respirometry measurements used in ASTM and ISO standards for biodegradation in marine environments. In aerobic conditions, the minimal amount of plastic samples that must biodegrade to carbon dioxide might range from 60 to 70% for three months for ASTM D6691-09, six months for ASTM D7473-12, or twenty-four months for ISO 18830, ISO 19679, and ASTM D7991-15 [3]. When organic matter is absorbed by polymers, carbon dioxide (CO_2) may be generated on the sample or absorbed by it [3]. One way to verify biodegradation would be as follows:

- Taking into account variations in bulk, appearance, and mechanical characteristics.
- Taking into account combined modifications to the molecular structure of the polymer, such as surface hydrolysis measurements using spectrophotometers.
- Examining the effects of microbes and/or biofilm formation using optical, atomic force, and scanning electron microscopy.

Although they are not specified in current standards, existing standards occasionally recommend additional respirometry measurements for confirmation [3]. The biodegradation of organic matter in the ocean would be restricted by nitrogen, which would also prevent the amounts of biodegradation claimed by published methods. Because bacteria require a limiting amount of nitrogen, the estimate for biodegradation using ASTM methods may be 6000 times higher than what may happen in the ocean. Additionally, the biodegradation time frame generated by the testing may be overestimated. As a result, the actual biodegradation of materials in maritime environments is probably exaggerated.

4.2   Plastic-Degrading Microbial Strains Isolated from the Alpine and Arctic

Actinobacteria, Proteobacteria, and Ascomycota are dominant taxa in the plastisphere of terrestrial cryo environments. The phyla Actinobacteria, Proteobacteria, and Ascomycota comprise nearly all of the microbial strains that were identified from the plastisphere of cold terrestrial habitats. The genus *Pseudogymnoascus* and the orders Burkholderiales, Pseudomonadales, and Streptomycetales were particularly frequently isolated, which is consistent with earlier reports of the soil plastisphere microbiome examined by microbial cultivation [8] and culture-independent genomic methods [9]. Moreover, the primary taxa known to be involved in the breakdown of PUR, PBAT, and PLA are these phyla (+Firmicutes) [10].

4.3   Cold-Adapted Bacteria and Fungi Are Able to Degrade Dispersed PUR and Polyester Films

The microbial strains from alpine and Arctic habitats to degrade dispersed PUR; and PBAT- and PLA-based polyester films at low temperatures (15 °C) has been noted and it was also observed that the maximum enzymatic activity of psychrophilic and psychrotolerant bacteria is above their upper growth limit, despite the fact that cold-adapted microbial strains frequently flourish at temperatures below 15 °C [11]. Further research is necessary to determine the ideal temperature for growth, enzyme production, and enzyme activity—all of which are important elements in the decomposition of plastic. More than half of the examined isolates have the capacity to break down the PUR dispersion Impranil®, according to our agar plate clearing experiment. As reported the Impranil® degraders made up about 30 % of the fungus identified from plastic debris floating in a lake [12]. Using agar clearing experiments with polycaprolactone, polybutylene succinate, and polybutylene succinateco-adipate at 28 °C, [13], it was demonstrated that around 39% of microbial strains obtained from Arctic habitats degraded at least one of the tested biodegradable polymers. The PUR was effectively degraded at low temperatures (15 °C) by members of the Ascomycota in particular. Barratt et al. [14] found that fungi were the main cause of PUR degradation in soil, which is consistent with our findings. The taxa *Rhodococcus* [15], *Pseudogymnoascus* [8], *Pseudomonas* [16], *Penicillium* [12], and *Verticillium* [17] have all been shown to degrade PUR. It seems to be the first instance of Impranil® being degraded by the fungal genera *Lachnellula, Neodevriesia,* and *Thelebolus,* as well as the bacterial genera *Amycolatopsis, Collimonas, Kribbella, Psychrobacter,* and *Streptomyces.* Additionally, five microbial strains could degrade BI-OPL at 15 °C and that 12 strains could significantly break down ecovio®. These two commercial polymers are composed of PLA and PBAT together with additional, unidentified ingredients and additives. Therefore, based alone on complete weight loss, it cannot be assumed that one of those polymers has been depolymerized. However, since commercial plastics are really released into the environment and recycled instead of pure polymers, studying the degradation of commercial plastic products is pertinent. Therefore, in order to be beneficial for remediation and recycling applications, appropriate microbes and enzymes would need to be able to degrade solid, commercial products that contain synthetic polymer components together with other components and additives. It's interesting to note that polycaprolactone has so far been the main target of breakdown by cold-adapted bacteria [18]. For example, when polycaprolactone films were incubated with an Arctic *Clonostachys rosea* strain for one month at 21 °C, [13] reported a 34.5% weight loss. Furthermore, according to the same authors, a *Pseudogymnoascus* strain from Antarctic soil degraded polycaprolactone (6%) and poly(butylene succinate-co-butylene adipate) (26 %) in 1 month at 14 °C [19]. This is similar to some of our strains using ecovio® and BI-OPL films at 15 °C. Thus far, only mesophilic and thermophilic microbial strains have been discovered to degrade PLA and PBAT. Leptothrix, a taxon not isolated in our investigation, was found to degrade ecoflex® (PBAT) plastic sheets at 30 °C, according to [20]. Another study found that after 20 days of incubation with *Rhodococcus fascians* at 25 °C, PBAT films lost 9% of their weight [21]. A strain of the same genus (958 *Rhodococcus sovatensis*) in our investigation was able to hydrolyze PBAT in the 4-MUL assay and degrade Impranil®, but it did not considerably lower the weight of the plastic sheets. The phylum Actinobacteria, which includes the genera *Amycolatopsis, Streptomyces,* and *Saccharothrix,* as well as temperatures of 30 °C or higher, have been shown to degrade PLA in prior research [22]. In our investigation, the weight of ecovio® and BI-OPL films was considerably decreased by strains 964 (*Streptomyces sp.*) and 985 (*Amycolatopsis sp.*). 985 (*Amycolatopsis sp.*) was able to break down the PBAT and PLA components of the BI-OPL film, while 964 (*Streptomyces sp.*) could only break down ecovio® components other than PBAT and PLA. According to our findings, these genera members may also be able to break down plastics at low temperatures (15 °C). Umezawaea tangerina, or strain 762, considerably decreased ecovio®'s weight. It considerably weakened the PLA component of the polymeric films when cultivated in R2A. Although Umezawaea is closely related to the PLA-degrading genus *Saccharothrix*, it has not been demonstrated to digest polymers before [23]. In a previous study, we also discovered enrichment of the genus in the plastisphere of biodegradable plastics in alpine soil, even though strain 762 was isolated from plastic collected in Svalbard [24]. It has been demonstrated that the plastisphere of biodegradable plastics in alpine and Arctic soils is enriched in the fungus species *Pseudogymnoascus*, which is prevalent in cold climate soils [24]. The weight of ecovio® was considerably decreased by every *Pseudogymnoascus* isolate that was tested here. It's interesting to note that while the sole strain of *P. roseus* (strain 967) was unable of degrading Impranil®, strains of the *Pseudogymnoascus* species *P. pannorum* and *P. verrucosus* were able to do so. Additionally, the PLA and PBAT components of ecovio® were not broken down by *P. roseus*. The PLA (1,031 *P. verrucosus*) and PBAT (1,034 and 1,205 *P. pannorum*, 966 and 1,031 *P. verrucosus*) components of the ecovio® films were broken down by other *Pseudogymnoascus* strains. According to a recent study by [19], an Antarctic *Pseudogymnoascus* strain degraded polycaprolactone and poly(butylene succinate-co-butylene adipate) at 14 °C. Interestingly, this strain degraded plastic more quickly at 14 °C than at 20 °C. They discovered that the *Pseudogymnoascus* strain produces two enzymes that can break down a variety of biodegradable polyesters [25]. When combined, these findings show that the genus *Pseudogymnoascus* has a broad capacity to break down polyester-type polymers at low temperatures. At least one of the plastic films could be broken down by the fungus strains 737 (*Thelebolus globosus*), 918 (*Penicillium stoloniferum*), 1,207 (*Oidiodendron echinulatum*), 800 (*Neodevriesia sp.*), and 943 (*Lachnellula sp.*). In addition to drastically degrading the PBAT and PLA components of the plastic films, the final two strains were able to lower the weight of both ecovio® and BI-OPL. In BI-OPL films, strain 737 (*Thelebolus globosus*) considerably reduced the PBAT and PLA when cultivated in R2A and the PBAT when cultivated in MM. According to [26], the psychrophilic fungus genus *Thelebolus* is known to flourish in Antarctic lakes. Its adaptation to cold conditions is highlighted by the fact that it has been demonstrated to produce extracellular enzymes (α-amylase) with a maximum activity at 20 °C and quickly diminishing activity at higher temperatures [27]. Both saprotrophic and plant-pathogenic species are found in Lachnellula, which can lead to diseases like larch canker [28]. Because plant-pathogenic species can manufacture cutinases that target polyesters because they resemble the plant polymeric component cutin, it has been frequently reported that these species can degrade polyesters [29]. This is the first time that strains of *Thelebolus* and *Lachnellula* have been reported in relation to plastic breakdown, as far as we are aware. Numerous environments are home to the genus *Neodevriesia* [30]. A strain of *Neodevriesia* was recently identified from a seashore that can generate halos on agar that contains polycaprolactone [31]. It has been discovered that the plastisphere of biodegradable plastics in alpine soil contains a higher concentration of the genus *Oidiodendron* [24]. It has been demonstrated that certain species in this genus, which primarily live in soil and decomposing plant matter, produce enzymes such as lipases, gelatinases, and polyphenol oxidases. Additionally, *Oidiodendron* has been linked to the degradation of both synthetic and natural rubber [32]. Strains of *Penicillium* are renowned for their wide range of metabolic activities, which include breaking down different kinds of plastic [33]. By generating plastic-degrading enzymes that are active at lower temperatures, the isolated cold-adapted microbes may also be helpful for recycling or upcycling in addition to their value for the environmental break-down of plastic. It should be noted, nevertheless, that the strains that were examined took a considerable time to break down the plastic films (analyzed after 60 and 126 days). Furthermore, strain 1,205 (*Pseudogymnoascus pannorum*) for ecovio® and strain 737 (*Thelebolus globosus*) for BI-OPL achieved the greatest weight loss in our investigation after 60 days, compared to earlier studies that showed nearly total disintegration of plastic films [34]. Inadequate incubation times, the incapacity to break down particular film com-ponents, and the material characteristics of the films (such as crystallinity) could all be contributing factors to the partial degradation. The crystalline domains of semi-crystalline polymers, such PBAT, are more resistant than the amorphous domains [35]. It's inter-esting to note that we found a significant linear relationship between the mass loss of PBAT and the amount of terephthalate in both biodegradable plastic films. This suggests that enzymes broke down the films because they preferentially target the butylene adipate-rich components of PBAT. The insufficient decomposition of plastic films in our tests may potentially be due to the higher ratios of aromatic to aliphatic groups in PBAT, which have been demonstrated to slow down hydrolysis by enzymes [36]. Other theories include waste product buildup, nutrient depletion, and acidity of the culture media, all of which may eventually prevent microbial growth and plastic decomposition. Not surprisingly, over the tested period, we were unable to identify any microbial strains that could break down PE. Only a small number of studies showed that microbial strains could alter the material properties of untreated PE, and the majority only partially degraded artificially weathered PE (e.g., by UV radiation) [37]. PE can be oxidized by such pretreatments, increasing its biodegradability [38]. PE is generally regarded as non-biodegradable in environmental circumstances, despite certain studies demonstrating signs of microbial breakdown.

4.4   Factors Affecting Plastic Biodegradation

The properties of the polymer, the exposure circumstances, and the enzymes all have an impact on the biodegradation process. Below is a list of some of these factors.

Exposure Conditions

Moisture
Because water is necessary for microbial growth and multiplication, moisture can affect polymer biodegradation in a variety of ways. Therefore, rapid microbial action increases the rate of polymer degradation when there is adequate moisture present [39]. Addition-ally, by producing additional chain scission reactions, moisture-rich circumstances aid in the hydrolysis process.

PH and Temperature
By altering the acidic or basic conditions, the pH can change the pace of hydrolysis reactions. For instance, PLA capsules hydrolyse well at a pH of 5 [40]. Different polymer degradation products change the pH levels, which in turn affect the rate of microbial development and degradation. Similarly, the softening temperature of the polymer has a major impact on enzymatic degradability. The likelihood of biodegradation is lower for polyester with a higher melting point. As the temperature rises, potential enzymatic degradability falls. For example, *R. delemar's* purified lipase effectively hydrolyzed polyesters with low melting points, such as PCL [41].

Enzyme Characteristics
Different enzymes can biodegrade different kinds of polymers and each has its own active site. For example, compared to straight chain polyesters made from any other monomer, those made from diacid monomers containing six to twelve C-atoms have been rapidly broken down by enzymes made by the fungus species *A. flavus* and *A. niger* [42]. It was discovered that the extracellular enzymes (depolymerases) that break down PHB through different processes rely on the particular microbially generated depolymerase [43]. Because of their hydrophobicity and three-dimensional structure, plastics made from petrochemical sources are difficult for the environment to break down [43]. Furthermore, PE's hydrophobic properties prevent bacteria from forming a biofilm, slowing down the pace of biodegradation [44].

Polymer Characteristics

Molecular Weight
From the perspective of biodegradability, molecular weight is crucial in determining a variety of polymer characteristics. As molecular weight increases, degradability decreases. Compared to low molecular weight polymers, lipase of a strain *R. delemar* decomposed higher molecular weight PCL (>4000) more slowly [35]. Microbial enzymes can more easily target a substrate with a low molecular weight [45].

Shape and Size
The polymer's characteristics, such as its size and form, are crucial to the degrading process. Compared to polymers with a limited surface area, those with a big surface area can deteriorate more quickly. For the biodegradation of different types of plastics, there is a common criterion of size and form [46].

Additives
The degradation ability is impacted by non-polymeric impurities like filler or dyes, which are waste or debris of catalysts used for polymerization and additive conversion products. It has been said that an increase in lingo-cellulosic filler causes the sample's thermal stability to decrease, which is followed by an increase in the amount of ash. The primary

determinants of the composite system's thermal stability are the lingo-cellulosic filler's dispersion and interfacial adhesion with the thermoplastic polymer [47]. In the production of polyolefins from polymers susceptible to thermo-oxidative breakdown, metals also function well as pro-oxidants.

Biosurfactants

Amphiphilic substances called biosurfactants are mostly generated on the surfaces of living things. Because biosurfactants are highly biodegradable and have minimal toxicity, they improve the biodegradation of both fossil and bio-based polymers. Because they include particular functional groups that aid in the biodegradation process, biosurfactants enable action under situations of high salinity, pH, and temperature [48].

4.5 Suggestions and Recommendations

Because of their widespread use and direct outdoor emission, plastics are currently present in our environment at dangerously high concentrations. To remove plastics from the environment and preserve living things, a variety of physicochemical techniques are employed, including photooxidative, thermal, ozone, mechanochemical, and catalytic. However, these techniques are expensive and unsuitable for usage at low plastic concentrations. Comparing microbial use to conventional methods for plastics decomposition, the former is now thought to be more environmentally benign. Below are some recommendations to take into account while using microbes to prevent plastic breakdown.

a. To degrade plastic potential of microbes including bacteria, fungi, and algae should be investigated.
b. To maintain optimum conditions of microbes for efficient plastic extermination.
c. Use of appropriate consortium of aerobic and anaerobic bacteria for more efficient plastic degradation.
d. Successive use of microbes (bacteria, fungi, and algae) can also be effective for plastic degradation.
e. Use of microbial enzymes, e.g., laccase, lignin degrading enzymes, urease, lipase, and protease can also be exploited to degrade plastic under aerobic and anaerobic conditions.

Thus, microbial consortia, their mechanisms, and their enzymes can be used to degrade plastic in a sustainable way.

References

1. UNEP, Biodegradable Plastics and Marine Litter. Misconceptions, Concerns and Impacts on Marine Environments, United Nations Environment Programme (UNEP), Nairobi, 2015.
2. UNEP, Marine Plastic Debris and Microplastics – Global Lessons and Research to Inspire Action and Guide Policy Change, United Nations Environment Programme, Nairobi, 2016.
3. Jacquin Justine, Jingguang Cheng, Odobel Charlene, Pandin Caroline, Conan Pascal, Pujo-Pay Mireille, Barbe Valerie, Meistertzheim Anne-Leila, Ghiglione Jean-François, Microbial ecotoxicology of marine plastic debris: a review on colonization and biodegradation by the "plastisphere", Frontiers in Microbiology Journal 10 (2019). https://www.frontiersin.org/articles/ https://doi.org/10.3389/fmicb.2019.00865/full. (Accessed 5 October 2020).
4. A.C. Martiny, et al., Concentrations and ratios of particulate organic carbon, nitrogen, and phosphorus in the global ocean, Sci. Data 1 (2014) 140048, https://doi.org/10.1038/sdata.2014.48. https://www.ncbi.nlm.nih.gov/pmc/articles/PMC4421931/pdf/sdata201448.pdf. (Accessed 25 May 2021).
5. M. Voss, H.W. Bange, J.W. Dippner, J.J. Middelburg, J.P. Montoya, B. Ward, The marine nitrogen cycle: recent discoveries, uncertainties and the potential relevance of climate change, Phil Trans R Soc B 368 (2013) 20130121, https://doi.org/10.1098/rstb.2013.0121. (Accessed 9 October 2020).
6. J. Masura, et al., Laboratory Methods for the Analysis of Microplastics in the Marine Environment: Recommendations for Quantifying Synthetic Particles in Waters and Sediments, vol. 48, NOAA Technical Memorandum NOS-OR&R, 2015.
7. Amir Reza Bagheri, Christian Laforsch, Andreas greiner,* and seema agarwal*, fate of so-called biodegradable polymers in seawater and freshwater, Global Challenges 1 (2017) 17000481. https://onlinelibrary.wiley.com/doi/pdf/https://doi.org/10.1002/gch2.201700048. (Accessed 30 May 2021).
8. Cosgrove, L., McGeechan, P. L., Robson, G. D., and Handley, P. S. (2007). Fungal communities associated with degradation of polyester polyurethane in soil. Appl. Environ. Microbiol. 73, 5817–5824. https://doi.org/10.1128/AEM.01083-07.
9. Zhang, K., Hamidian, A. H., Tubić, A., Zhang, Y., Fang, J. K. H., Wu, C., et al. (2021). Understanding plastic degradation and microplastic formation in the environment: a review. Environ. Pollut. 274:116554. https://doi.org/10.1016/j.envpol.2021.116554.
10. Gambarini, V., Pantos, O., Kingsbury, J. M., Weaver, L., Handley, K. M., and Lear, G. (2021). Phylogenetic distribution of plastic-degrading microorganisms. Microb Syst 6, e01112–e01120. https://doi.org/10.1128/mSystems.01112-20.
11. Huston, A. L., Krieger-Brockett, B. B., and Deming, J. W. (2000). Remarkably low temperature optima for extracellular enzyme activity from Arctic bacteria and sea ice. Environ. Microbiol. 2, 383–388. https://doi.org/10.1046/j.1462-2920.2000.00118.x.
12. Brunner, I., Fischer, M., Rüthi, J., Stierli, B., and Frey, B. (2018). Ability of fungi isolated from plastic debris floating in the shoreline of a lake to degrade plastics. PLoS One 13:e0202047. https://doi.org/10.1371/journal.pone.0202047.
13. Urbanek, A. K., Rymowicz, W., Strzelecki, M. C., Kociuba, W., Franczak, Ł., and Mirończuk, A. M. (2017). Isolation and characterization of Arctic microorganisms decomposing bioplastics. AMB Express 7:148. https://doi.org/10.1186/s13568-017-0448-4.
14. Barratt, S. R., Ennos, A. R., Greenhalgh, M., Robson, G. D., and Handley, P. S. (2003). Fungi are the predominant micro-organisms responsible for degradation of soil-buried polyester polyurethane over a range of soil water holding capacities. J. Appl. Microbiol. 95, 78–85. https://doi.org/10.1046/j.1365-2672.2003.01961.x.

15. Akutsu-Shigeno, Y., Adachi, Y., Yamada, C., Toyoshima, K., Nomura, N., Uchiyama, H., et al. (2006). Isolation of a bacterium that degrades urethane compounds and characterization of its urethane hydrolase. Appl. Microbiol. Biotechnol. 70, 422–429. https://doi.org/10.1007/s00253-005-0071-1.

16. Cregut, M., Bedas, M., Durand, M. J., and Thouand, G. (2013). New insights into polyurethane biodegradation and realistic prospects for the development of a sustainable waste recycling process. Biotechnol. Adv. 31, 1634–1647. https://doi.org/10.1016/j.biotechadv.2013.08.011.

17. Navarro, D., Chaduli, D., Taussac, S., Lesage-Meessen, L., Grisel, S., Haon, M., et al. (2021). Large-scale phenotyping of 1,000 fungal strains for the degradation of non- natural, industrial compounds. Commun. Biol. 4, 871–810. 10.1038/ s42003-021-02401-w.

18. Sekiguchi, T., Saika, A., Nomura, K., Watanabe, T., Watanabe, T., Fujimoto, Y., et al. (2011). Biodegradation of aliphatic polyesters soaked in deep seawaters and isolation of poly(ε-caprolactone)-degrading bacteria. Polym. Degrad. Stab. 96, 1397–1403. https://doi.org/10.1016/j.polymdegradstab.2011.03.004.

19. Urbanek, A. K., Strzelecki, M. C., and Mirończuk, A. M. (2021). The potential of coldadapted microorganisms for biodegradation of bioplastics. Waste Manag. 119, 72–81. https://doi.org/10.1016/j.wasman.2020.09.031.

20. Nakajima-Kambe, T., Toyoshima, K., Saito, C., Takaguchi, H., Akutsu-Shigeno, Y., Sato, M., et al. (2009). Rapid monomerization of poly(butylene succinate)-co-(butylene adipate) by Leptothrix sp. J. Biosci. Bioeng. 108, 513–516. 10.1016/j. jbiosc.2009.05.018.

21. Soulenthone, P., Tachibana, Y., Muroi, F., Suzuki, M., Ishii, N., Ohta, Y., et al. (2020). Characterization of a mesophilic actinobacteria that degrades poly(butylene adipate-coterephthalate). Polym. Degrad. Stab. 181:109335. https://doi.org/10.1016/j.polymdegradstab. 2020.109335.

22. Sriyapai, P., Chansiri, K., and Sriyapai, T. (2018). Isolation and characterization of polyester-based plastics-degrading bacteria from compost soils. Microbiol (Rus. Fed.) 87, 290–300. https://doi.org/10.1134/S0026261718020157.

23. Labeda, D. P., and Kroppenstedt, R. M. (2007). Proposal of Umezawaea gen. nov., a new genus of the Actinosynnemataceae related to Saccharothrix, and transfer of Saccharothrix tangerines Kinoshita et al. 2000 as Umezawaea tangerina gen. nov., comb. nov. Int. J. Syst. Evol. Microbiol. 57, 2758–2761. https://doi.org/10.1099/ijs.0.64985-0.

24. Rüthi, J., Bölsterli, D., Pardi-Comensoli, L., Brunner, I., and Frey, B. (2020). The "Plastisphere" of biodegradable plastics is characterized by specific microbial taxa of alpine and Arctic soils. Front. Environ. Sci. 8:562263. https://doi.org/10.3389/fenvs.2020.562263.

25. Urbanek, A. K., Arroyo, M., Mata, I.De, and Mirończuk, A. M. (2022). Identification of novel extracellular putative chitinase and hydrolase from Geomyces sp. B10I with the biodegradation activity towards polyesters. AMB Express, 12:12 https://doi.org/10.1186/s13568-022-01352-7.

26. de Hoog, G. S., Göttlich, E., Platas, G., Genilloud, O., Leotta, G., and van Brummelen, J. (2005). Evolution, taxonomy, and ecology of the genus Thelebolus in Antarctica. Stud. Mycol. 51, 33–76.

27. Singh, S., Singh, P., Singh, S., and Sharma, P. (2014). Pigment, fatty acid and extracellular enzyme analysis of a fungal strain Thelebolus microsporus from Larsemann Hills, Antarctica. Polar Rec. 50, 31–36. https://doi.org/10.1017/S0032247412000563.

28. Giroux, E., and Bilodeau, G. J. (2020). Whole genome sequencing resource of the European larch canker pathogen Lachnellula willkommii for molecular diagnostic marker development. Phytopathology 110, 1255–1259. 10.1094/ PHYTO-09-19-0350-A.

29. Brodhagen, M., Peyron, M., Miles, C., and Inglis, D. A. (2015). Biodegradable plastic agricultural mulches and key features of microbial degradation. Appl. Microbiol. Biotechnol. 99, 1039–1056. https://doi.org/10.1007/s00253-014-6267-5.

30. Wang, M.-M., Shenoy, B. D., Li, W., and Cai, L. (2017). Molecular phylogeny of Neodevriesia, with two new species and several new combinations. Mycologia 109, 965–974. https://doi.org/10.1080/00275514.2017.1415075.
31. Kim, S. H., Lee, J. W., Kim, J. S., Lee, W., Park, M. S., and Lim, Y. W. (2022). Plasticinhabiting fungi in marine environments and PCL degradation activity. Antonie Van Leeuwenhoek 115, 1379–1392. https://doi.org/10.1007/s10482-022-01782.
32. Lugauskas, A., Prosychevas, I., Levinskaitė, L., and Jaskelevičius, B. (2021). Physical and chemical aspects of long-term biodeterioration of some polymers and composites. Environ. Toxicol. 19, 318–328. https://doi.org/10.1002/tox.20028.
33. Srikanth, M., Sandeep, T. S. R. S., Sucharitha, K., and Godi, S. (2022). Biodegradation of plastic polymers by fungi: a brief review. Bioresour. Bioprocess. 9:42. 10.1186/ s40643-022-00532-4.
34. Tokiwa, Y., and Jarerat, A. (2005). Accelerated microbial degradation of poly(Llactide). Macromol. Symp. 224, 367–376. https://doi.org/10.1002/masy.200550632.
35. Tokiwa Y, Calabia BP, Ugwu CU, Aiba S (2009) Biodegradability of plastics. Int J Mol Sci 10(9):3722–3742. https://doi.org/10.3390/ijms10093722.
36. Zumstein, M. T., Rechsteiner, D., Roduner, N., Perz, V., Ribitsch, D., Guebitz, G. M., et al. (2017). Enzymatic hydrolysis of polyester thin films at the nanoscale: effects of polyester structure and enzyme active-site accessibility. Environ. Sci. Technol. 51, 7476–7485. https://doi.org/10.1021/acs.est.7b01330.
37. Sowmya, H. V., Ramalingappa, K., Krishnappa, M., and Thippeswamy, B. (2015). Degradation of polyethylene by Penicillium simplicissimum isolated from local dumpsite of Shivamogga district. Environ. Dev. Sustain. 17, 731–745. 10.1007/ s10668-014-9571-4.
38. Mohanan, N., Montazer, Z., Sharma, P. K., and Levin, D. B. (2020). Microbial and enzymatic degradation of synthetic plastics. Front. Microbiol. 11:580709. 10.3389/ fmicb.2020.580709.
39. Ho K-LG, Pometto AL III, Hinz PN (1999) Effects of temperature and relative humidity on polylactic acid plastic degradation. J Environ Polym Degrad 7(2):83–92. https://doi.org/10.1023/A:1021808317416.
40. Henton DE, Gruber P, Lunt J, Randall J (2005) Polylactic acid technology natural fibers, biopolymers, and biocomposites 16:527–577.
41. Tokiwa Y, Calabia BP (2004) Review degradation of microbial polyesters. Biotechnol Lett 26(15):1181–1189. https://doi.org/10.1023/B:BILE.0000036599.15302.e5.
42. Kale G, Kijchavengkul T, Auras R, Rubino M, Selke SE, Singh SP (2007) Compostability of bioplastic packaging materials: an overview. Macromol Biosci 7(3):255–277. https://doi.org/10.1002/mabi.200600168.
43. Yamada-Onodera K, Mukumoto H, Katsuyaya Y, Saiganji A, Tani Y (2001) Degradation of polyethylene by a fungus, Penicillium simplicissimum YK. Polym Degrad Stab 72(2):323–327. https://doi.org/10.1016/S0141-3910(01)00027-1.
44. Hadad D, Geresh S, Sivan A (2005) Biodegradation of polyethylene by the thermophilic bacterium Brevibacillus borstelensis. J Appl Microbiol 98:1093–1100.
45. Auras R, Harte B, Selke S (2004) An overview of polylactides as packaging materials. Macromol Biosci 4(9):835–864. https://doi.org/10.1002/mabi.200400043.
46. Kijchavengkul T, Auras R (2008) Compostability of polymers. Polym Int 57(6):793–804. https://doi.org/10.1002/pi.2420.
47. Yang H-S, Wolcott M, Kim H-S, Kim H-J (2005) Thermal properties of lignocellulosic filler-thermoplastic polymer bio-composites. J Therm Anal Calorim 82(1):157–160. https://doi.org/10.1007/s10973-005-0857-5.
48. Kawai F, Watanabe M, Shibata M, Yokoyama S, Sudate Y (2002) Experimental analysis and numerical simulation for biodegradability of polyethylene. Polym Degrad Stab 76(1):129–135. https://doi.org/10.1016/S0141-3910(02)00006-X.